SCIENTIFIC AND MEDICAL KNOWLEDGE PRODUCTION, 1796–1918

SCIENTIFIC AND MEDICAL KNOWLEDGE PRODUCTION, 1796–1918

Experiment, Expertise, Experience

Edited by
Rob Boddice

Volume I,
Curiosity

Routledge
Taylor & Francis Group
LONDON AND NEW YORK

First published 2023
by Routledge
4 Park Square, Milton Park, Abingdon, Oxon OX14 4RN

and by Routledge
605 Third Avenue, New York, NY 10158

Routledge is an imprint of the Taylor & Francis Group, an informa business

British Library Cataloguing-in-Publication Data
A catalogue record for this book is available from the British Library

ISBN: 978-0-367-44373-3 (set)
ISBN: 978-0-367-44374-0 (Volume I) hbk
ISBN: 978-1-003-00933-7 (Volume I) ebk

DOI: 10.4324/9781003009337

Typeset in Times New Roman
by Apex CoVantage, LLC

CONTENTS

ACKNOWLEDGEMENTS

This collection has been a trial by fire, water and disease. I wish I were speaking metaphorically, but I am not. Thanks first and foremost go to Mary Yearl, librarian at the Osler Library for the History of Medicine at McGill University, who managed to unearth my requests for material in the midst of a librarian's worst nightmare. Thankfully, both the library and this project were saved from what could have been an unmitigated disaster. My thanks also go to McGill's digitisation and delivery services, which have rendered so many of these rare materials easy to access and use, and to Mary and the Osler Library for permission to reproduce them in this book.

The McGill materials make up about a quarter of the finished project. For the rest, I am indebted to the brilliant assistance of Marie van Haaster, who very quickly understood what the project entailed and equally rapidly brought herself up to speed on the subject matter. Working with a broad brief and thematic way finders, she effectively curated a digital archive of material from which to draw the key sources. Organising those sources and the Osler material into coherent volumes has been a particular challenge, but it would have been impossible without these two key sustained acts of assistance.

Thanks go to Rachel Douglas for inspiring this idea and getting it off the ground. I am indebted also to the directors of the Academy of Finland Centre of Excellence in the History of Experiences, Tampere University, Finland, for the time and freedom to pursue such a grand and unwieldy project.

GENERAL INTRODUCTION
Experience, Experiment, Expertise

Nineteenth-century scientists were much more comfortable with the literal meaning of 'scientist' than many who currently practice under that term. Just as the *art-ist* is someone who makes art, so the *scient-ist* literally makes knowledge. The knowledge acquired through scientific practices is not found, discovered or unveiled. It is *made, constructed, produced.*[1] Sometimes, it is produced in the cultural idiom of finding, discovering or unveiling; sometimes it is produced in the cultural idiom of neutrality, objectivity and control. But these things are cultural applications to scientific knowledge production (both the process and the results); they do not actually make scientific knowledge neutral or objective or controlled in some timeless way.

I suspect that a large community of scientists would be uncomfortable with the description of scientific practice as a cultural activity, and of scientific knowledge as a product of such activity. Scientists have long persevered in their resistance to being studied by social scientists and others. But the situatedness of scientific knowledge production is the key observation of the history of science and of science studies, which document not a teleological progression of ideas that ends at either a current state of enlightened perfection or else tends in that direction. On the contrary, the history of science documents a messy assemblage of faulty ideas and failures, often dressed up in their own contemporary garb of righteousness and factual certainty, undergirding social and political decision-making, their wrongness notwithstanding. The history of science shows the cultured structure and politics of scientific-knowledge production, through its situated practices, which are often disciplinarily specific, forming a contained 'epistemic culture', *Denkkollektiv*, or 'moral economy'.[2] In whatever way contemporary scientists might differ in their knowledge claims and in their methods, they are not exempt from a cultural and political situation. Knowledge is produced, inescapably, through such structures.[3] Protestations to the contrary are no less situated. There is no way out of one's context.

This being the case, what are the specific ingredients that account for things like certainty, or the expression of discovery as opposed to construction? In what does scientific authority inhere? What drives the processes of knowledge production? And how, in contexts of certainty and amid claims of objectivity, does knowledge change?

DOI: 10.4324/9781003009337-1

What happens when certainty is shaken? What is the situated value of uncertainty among scientists, and what does it do to the status of the knowledge they produce?

This collection of sources is about all these fundamental guiding questions with regard to the production of scientific knowledge production. I have situated them within a thematic concordance of three key categories – experience, experiment, expertise – but it is perhaps easier to think of the single master category here, from which all these terms derive: *experior*. The Latin roots capture the quality of experience derived from *doing* or *attempting*, where the practice of experiment is the experience that produces the expertise. In literal terms, there is no expertise without experience and there is no experience without experiment. To situate this in a yet older form, in science one needed ἐμπειρίᾱ, where knowledge came 'out of' the *trial* or the attempt. From this we get the English word 'empirical', which also captures the essence of science as a practice, as a verb, as something *done*. Knowledge comes out of this doing.

My reason for approaching the subject through the compound meanings of *experior* are justified by the texts themselves. Through these sources, I am following the development across many different disciplines of scientific endeavour to enshrine experimentation at the heart of scientific knowledge claims. This coincides with a process of specialisation and professionalisation that the emphasis on experiment seemed to encourage. While natural philosophers and gentleman scholars might have wielded insights across a range of topics, the emphasis on *doing* in the name of science gradually eroded the value of knowledge derived from textual learning.[4] Reading accumulations of facts in books could not hold against the expertise gained from recreating knowledge-production practices for oneself. The very status of knowledge was at stake. For at the core of the impetus to experiment for oneself, or the exhortation to others that they should experiment for themselves, was the breach of trust in the authority of the word and of inherited tradition. Knowledge, in scientific modernity, had to be tested and reproduced in order to be safeguarded. As ancient certainties came unstuck, in part through mathematical innovation, and in part through looking, from cosmological knowledge of the solar system and the wider universe to the anatomy and physiology of the human and animal body, so everything became the subject of renewed scrutiny.[5] To be believed, knowledge had to be seen. And to see meant to do. This allowed, on the one hand, for a great expansion of knowledge in multiple new specialisms, where the focus was microscopic or telescopic, or else *seeing* was displaced through the physical or chemical evidence: movement, points on a graph, isolation of bodily fluids or elements. On the other hand, it meant that the days of the great sage were numbered.

RATIONALE

The roots of these volumes lie in the research for three of my other books and in an exhibition that did not happen, for a variety of reasons. The three books are *Edward Jenner*; *The Science of Sympathy: Morality, Evolution, and Victorian*

Civilization; and *Humane Professions: The Defence of Experimental Medicine, 1876–1914*.[6] Jenner's biography, which was based on the diffuse manuscript and printed sources pertaining to his life and career, exemplifies both the processes of knowledge production within medical science and the political fortunes of that knowledge in the context of a resistant prevailing paradigm and pecuniary interests. Jenner was famed for having introduced 'control' into his experiments (Jenner's experimental records appear early in this collection), but there is as much blind fumbling as there is planned research in his experiments on children and animals with respect to both smallpox and tuberculosis. He is famed for his successful intervention in preventing the former, but his catastrophic failures in understanding the latter are generally unacknowledged. Even the success story, however, was, in his own lifetime, unsecured as accepted scientific truth. Despite awards from the public purse and a slew of honours, at the end of his life his own convictions about the efficacy of the vaccination he had pioneered were not universally shared, and it would take decades of political wrangling and modifications of vaccination methods before its status was secured in Britain. Even then, it took until 1978 for smallpox to be eradicated in Britain, and until 1980 for smallpox to be eradicated globally. The status of vaccination in general seems, in the present, to be on a less secure footing than ever. The status of scientific knowledge in the world is never reducible to the facts themselves. Facts are mobilised. Partial truths are represented as the whole truth or else dismissed as false. Methods declared safe sometimes prove to be unsafe, and while the bare fact, say, of vaccine efficacy remains unchanged, political wranglings over practices, ethics and individual freedoms seem to put the knowledge claims themselves in doubt. So much can be gleaned from Jenner's story, and from the fortunes of the vaccine in the remainder of the nineteenth century.

These I pursued in *The Science of Sympathy*, alongside the emergence of experimental physiology and the emergence of a statistical science of biometry and eugenics, in the context of intellectual experimentation with the principles of Charles Darwin's theory of evolution. That theory can itself be described as the product of experience, of empirical observation and study, and of experimental breeding (especially of plants, but also of animals, notably pigeons). It is also exemplary of the processes of establishing and securing scientific authority – a political as much as an empirical process – and of the application of knowledge claims in the name of humanity. In this case, the extension of knowledge claims into the realms of experimentation on the living, enforced vaccination under threat of legal penalty, and the scientific control of human breeding, were *all* put forward in the name of humane science, with scientists as the humane visionaries of social organisation and public health. Many of the sources I used in *Science of Sympathy* appear here as exemplary of the kinds of political claims that inhere in knowledge claims.

Humane Professions continues the theme of *Science of Sympathy*, but limits the focus, dealing with an international network of medical scientists and their allies in their defence of the practice of vivisection as a means of producing new knowledge and new medical therapies. Experimentation was the central thing at stake in this defence, which took the form of a cultural war over the meaning and practical implications of cognate terms like sympathy, compassion, pity, mercy, benevolence and humanity. Pitting themselves against a prevailing understanding of compassion in a Christian idiom of charity and pity, the scientific community pointed out that these feelings were often self-serving rather than of social benefit, and that real social good could only be done through rational and abstract means. In other words, the truly humane man (it was almost always a man) experimented on animals in his laboratory, producing knowledge that would, when properly applied, genuinely improve the lot of society, reducing its sufferings and maximising its potential for pleasures. Not only did this foreground the social importance of experimentation, but in emphasising the affective character of the men who carried out the experiments, it cut off their critics from the thing they wished to criticise. Charges of boundless and inhumane curiosity, of, at best, an inurement to the sight of blood or the howl of pain and, at worst, a real relish for those things, were thrown about. But the response to them was a wave of dismissal. How could anyone know how it felt to be an experimenter if they did not have first-hand experience of experimenting? What may look like blood lust, or of conscience abandoned to a compulsive curiosity, was actually sober, difficult work, undertaken with compunction, with the conviction that it would do *good*. Unless you were privy to the knowledge that undergirded this conviction (expertise), and had had a hand in either making or confirming this knowledge (experience), you had no business and no credibility in critiquing the means of making this knowledge (experiment). For all that scientists in this period promised to make their practices and premises open to public inspection and scrutiny, the internal logic of their culture of knowledge production meant that it could only be understood by scientists themselves. One could write down and publish the results, but from this one could only judge the outcome and implications of the knowledge produced. One could not judge the manner of its production.

Among the sources here assembled are some of the key documents that were the building blocks for that book. I had planned for those of them that are lodged in the Osler Library for the History of Medicine at McGill University in Montreal to be exhibited, with a narrative that looked a bit like this one, but the exhibition was postponed first by a fire and associated water damage and then by the Covid-19 pandemic. In some respects, these volumes have now become the exhibition, but it has become much more extensive than my initial plan would have allowed. I have tried to be as comprehensive as possible in my selection of sources, covering both the whole period of the nineteenth century in its broadest (longest) sense and a wealth of different scientific endeavours, from astronomy to chemistry to physiology to biology to psychology. Couched within these recognisable disciplinary tags – and this collection in many ways documents the processes of the emergence of these disciplines as recognisable – are subjects that do not, on first blush, seem to fit: mesmerism and spiritualism, for example. They are included because, in the context of their times, they absolutely staked a claim to inclusion. And the exhortation to derive proof through experimentation was as much theirs as it belonged to other sciences. Such broad applications of the value and practice of empiricism should give us pause, but I shall return to this topic presently.

For now, by way of a final first word, I simply wish to underscore the extent to which this collection has gone beyond the specific techniques and justifications of the laboratory, even though the laboratory idiom runs through even non-laboratory sciences like a red thread in this period. When pieced together, these sources document the fortunes of a well-rehearsed exhortation to test, to try, to repeat, to distrust the written record and produce and re-produce knowledge by doing. They amount to a story of experimental aggregation – of doing and re-doing, of the need for each neophyte to *go through* the same process of seeing it with his own eyes – that amounts to both individual and shared experience. To an extent these volumes amount to a kind of Rankean proof for the books I have written and for the position I have taken up, but in this respect they are partial – a sampling – and useless without the broader framework of scholarship in the field. While I introduce the rationale for this collection through my own experience of researching and writing three books on the history of science and medicine in the long nineteenth century, I want to underscore that these projects in turn represent an enormous body of scholarly literature that has been formative of my own work (I limit, where possible, the historiographical survey to the general introduction; the volume introductions are designed as general context for the sources that follow). I cannot hope to represent that corpus in its entirety here, but I do think it important to situate the reader, or perhaps the *user* is better, among the most important pillars of wisdom in the historiography. Again, it is only a starting point. Insofar as the primary sources are useful, they are useful only with this historiographical foundation, which in turn can unfold indefinitely. I advise the student entering into the field not to dispense with the secondary literature.

ORIENTATION

First a note about periodisation. To some extent, beginning and ending points are always arbitrary and always false. I do not want to make any strident claims that the year 1796 marks the *beginning* of any grand new structure or process, or that it marks the leaving behind of something decidedly 'unmodern'. The past reaches into the present and future at any moment, and this is certainly true of the life of my chosen starting point, Edward Jenner, a country doctor and God-fearing amateur generalist. His scientific world, of epistolary connections to a few great urban (though not, at first, urbane) names, of collecting, dissecting, pondering and tinkering, was thoroughly undisciplined, practically self-taught. It is easy, retrospectively, to isolate Jenner's fame to the discovery and distribution of knowledge about a safe way to prevent smallpox, but in Jenner's own lifetime that fame was much less secure. His first glory was a treatise on the migratory habits of the cuckoo, hitherto unknown, and he spent far more of his time tending to patients and trying, in vain, to understand tuberculosis. His vaccine mired him in politics and in a society from which he derived only suffering and discomfort.

His life was, by the time of his death in 1823, still very much characteristic of the eighteenth-century rural doctor on horseback, nosing in the hedgerows. But two things justify the starting point: first, a method, that showed through repeated experiments that an element of control could be introduced so as to provide at least an element of certainty. People had been getting their hands experimentally dirty for centuries, and Vesalius had dialled in the importance of empirical investigation, but Jenner was testing a hypothesis experimentally and then proving it experimentally, with repetition. It was a significant innovation. Second, Jenner offered his discovery to anyone who wanted it, for free. He would give the method and the material. It was a social, public-health initiative. It was met, by some of the medical establishment, with pecuniary scorn. It upended a culture that placed in the medical man a body of esoteric recipes and rituals inaccessible to the layman and therefore available only at a cost to the patient. Jenner's vaccine was thoroughly democratic: a medical procedure that anyone could carry out having read a one-page instruction leaflet. The willingness to give it away found its way into the centralisation of public-health policy and administration. By the 1850s, everyone was to have the vaccine, at no cost to them, under the direction of local administrators who in turn answered to the government. The public good, for some decades, mitigated the democratic sensibility, at least from a certain point of view. Vaccination became compulsory, with severe penalties for non-compliance. Many objected to the inroads into personal liberties, wondering where they might stop, but the very principle of the national distribution of a medical procedure for the sake of the common good of the whole population originated with Jenner.[7] This, at least, seems a suitable modern starting point for the connection between experience, experiment and expertise.

What about the end point? The year 1918 is a temporal convenience, coinciding with the end of the First World War, but there are a number of significant events in the years before this point that were suggestive of new turns and departures. The establishment of things like the Medical Research Council, itself the result of the implementation of the National Insurance Act of 1911, totally transformed the experimental landscape of medical science in Britain, bringing it under the umbrella of significant public funding.[8] The war itself, as documented by Todd Meyers and Stephanos Geroulanos, threw up countless experimental bodies, an unprecedented number of young men in various states of disintegration, transforming laboratory and clinical psychology, psychiatry, physiology, neurology and toxicology.[9] At the same time, those transformations asked serious questions of foregoing expertise, as well as driving these and other experimental sciences away from public scrutiny to the secretive world of government-funded military endeavours. In some ways, scientific authority and confidence in its expertise was at an all-time high by 1918, enshrined and funded at the core of governments across the world; in other ways, the war had knocked its confidence, rendered uncertain some of its certainties and tested its commitment to openness.

In the middle, there are certain key narratives that need to be addressed. The nineteenth century is the century of the emergence of the scientific professional, as well as the emergence of scientific and medical disciplines as institutional fixtures in places of higher learning.[10] It is the century of specialisation.[11] It is the century of evolution.[12] It is the century in which the foundations for a National Health Service were laid, even if they were not fully realised until much later.[13] It is the century of rising intellectual secularism, materialism and positivism, and therefore of a culture war among biologists, geologists and cosmologists alike about the meaning and content of the universe and the age of the earth.[14] It is the century of psychology, and of concrete sciences of emotion and madness.[15] It is the century of the laboratory.[16] It is also the century, somewhat paradoxically, of spiritualism, the ethereal, the mesmeric and the phrenological.[17] It is the era of the germ, the bacterium, the serum;[18] the era of antiseptic and anaesthetic.[19] It is the high age of industrialisation, drawing on and demanding more from physical and chemical engineers.[20] In these and many other ways, the nineteenth century does herald great shifts in experimentally founded scientific experience and expertise, as well as method, all of which had profound real-world consequences, socially, economically, imperially, culturally and medically.[21]

The histories of science and medicine have tended, over more than a generation, towards a mirroring of the processes of specialisation and disciplinary formation that took place within the nineteenth century, such that we have histories of physiology,[22] histories of astronomy,[23] histories of statistics[24] and race science,[25] sex science,[26] histories of chemistry[27] and physics,[28] and so on. It is by no means easy to put the historiography together in such a way as to represent the networked professional sociabilities that abided throughout this period, or to understand that the disciplinary and methodological logics of one field were connected, intellectually and culturally, to those of another.[29] Insight can be gained by a burgeoning biographical history of the period's leading scientific lights[30] and by an array of works that account for the rise of specific institutions of scientific learning and research,[31] the history of scientific instrumentation and technology,[32] and the formation and practice of gendered identity.[33] By juxtaposing a catholic literature, a certain thematic unity emerges that both justifies the selections in these volumes and, hopefully, inspires further study. The abundance of references that populate the preceding do not amount to a comprehensive literature survey, by any means, but for more of a rudimentary map. My overwhelming focus in these four volumes (and in this representation of the relevant literature) is on British sources – an artificial means of containing the sheer volume of material at my disposal – but such a limitation might have seemed arbitrary to the nineteenth-century scientist, and indeed many works here are English translations of works by prominent scientists working in other languages, whose fame was nonetheless secure on British shores, or works by American scientists whose fame was as bright on British shores. The American sources I include thematically intersect with prevailing issues of scientific knowledge production in Britain. So, while it is perfectly legitimate to study the history of science in a single polity or language, as here, it must be remembered that the themes explored in these volumes in an Anglophone context largely originated elsewhere, in Germany and France in particular, but also farther afield, and were carried to the new world as well as to the Empire.[34] These themes served colonial projects as much as they served claims of nationalist exceptionalism and independence. Any study of scientific experience and expertise must grapple with its globalised politics.[35]

In addition, I feel it necessary to reiterate my starting point and gesture toward the central importance in all these volumes of objectivity as a principle that served to guarantee methodological rigour, to underscore the validity of results and to serve as vehicle for scientific authority. Objectivity, both as method and as product, is at the core of nineteenth-century experimentalism. The very focus on empirical, observed evidence seemed to give to the scientist an authority that inhered in the experimental material rather than himself. As the mechanical means of seeing farther and/or better improved, so the appeal to the neutrality of technology augmented this appeal to the objectivity of scientific fact. This amounts to the great, institutionalised and popularised, self-deception of nineteenth-century scientists. It may have come across as self-effacing, is if nature unveiled herself under their watchful scrutiny, but it is clear that the appeal to objectivity was in fact a veil for a certain scripted culture of scientific knowledge production, in conformity with prevailing cultural and political ideologies. Objectivity, in Lorraine Daston and Peter Galison's words, is an 'affect' defined by its claim not to be an affect. It gives the researcher a certain mien, a certain bearing with respect to his work and to its presentation, that made a virtue out of looking, cutting, probing, testing and recording in certain ways, following certain logics, using certain technologies. The culture of objectivity within the moral economy of nineteenth-century science defined the claims to 'epistemic virtue' among its practitioners.[36] It was the well from which they drew their authority, and the source of their capacity to dismiss their opponents. Data, observation, statistics, the camera, the telescope, the microscope, the sphygmomanometer, the dolorimeter: a list of things, to which many others could be added, that are presented as incorruptible. But the data they produced requires a scientist to make it. The data has to be pushed through untested assumptions, cultural logics, methodological orthodoxies, argumentative styling, cosmological framing and political framing. Most of the time, these mitigating factors remain conveniently invisible, as if the journey from data to conclusion to implication had only one possible route, one possible outcome. Daston and Galison's book focuses on the development of the scientific 'atlas' as exemplary of this scientific cultural appeal to mechanical objectivity, but the central argument applies to every branch of experimental science in the nineteenth century.

SELECTION AND ORGANISATION

To a certain extent, each of these sources could appear in any of the four volumes, which eschew any attempt at a chronological scheme in favour of loose thematic and disciplinary divisions. The insistence on experimental practices of knowledge production occurs at slightly different moments in different fields of expertise, but there are certain structural similarities that tend to repeat. Each of the documents here could be read with a particular emphasis on one theme or another, such that I am aware that by so placing them in different sections there is a chance that I am overly directing the reader. This is not my intention, which is rather to highlight certain dominant notes across the disciplines, which underscore justifications for experimental practices at the expense of other methods, or else historically situate certain scientific practices with cultural and affective frameworks of interpretation.

The beginning and end points mirror each other. I begin with a focus on scientific curiosity. While on first blush one might infer from this theme an impetus to uncover what is unknown, the scientific rationale is more often situated in the idiom of material and/or positivist convictions that, while the desired knowledge is out of reach, the rationality of the method will *certainly* afford its production. The scientifically curious of the long nineteenth century often went through the motions of declaring that their endeavours could often end in failure or in surprise, but by far the more common claim was that properly guided (trained, dedicated, technologically and temperamentally suitable) curiosity would produce new and valuable truths. Whether such truths would reduce human and/or animal suffering, or heighten the understanding of the cosmos, or create some new industrial advantage, the raw benefit of the scientific bent – that is, the particular character of the scientist as the rational embodiment of curiosity – was an improvement in the lot of humanity: a universal good. Scientists were not wont to under-promise and over-deliver. Their scientism – a kind of deeply seated *faith* in science – led them, often, to promise the earth.

The reality, of course, is that many of their endeavours produced untruths that were, nonetheless, clothed in the garb of natural law. Curiosity, ultimately, *is* about what is unknown, but often the scientistic commitment to experimental processes and 'proofs' meant that many scientists in different fields were quick to assert the absolute finality of their findings, be it regarding vaccine efficacy or natural selection or the eugenic value of anthropometric data. If curiosity was often pursued in the pitch of hubris, a more sober admission of the need to revise, reform, retry and recast risked the status of scientific confidence. One can glimpse this wherever one looks in the history of science, including (and perhaps especially) in the present. But a great part of the experimental culture of scientific knowledge production lay in the fact that repetition and variation of experiments could challenge accepted truths, sometimes leading to rapid changes in understanding of a field. The mirror to the celebration of curiosity, therefore, is uncertainty as a guiding principle of research. The final volume thus places the emphasis on those experimental cultures that understood science as a process, and results as partial.

In between these two poles, the middle volumes focus on the political and cultural status of experimentation and scientific expertise. In particular, I have tried to bring together sources that showcase scientific claims to both the humanity of scientists and the humanitarianism of their work, such that expertise gains a moral framework. Where the ethics of experimentation, curiosity and a particular form of secular activity seemed a threat to more traditional, non-scientific understandings of morality and character, the capacity of scientists to put their work into a moral explanatory was, to them, of immense value. It not only justified what they did for their own sake, but it was the basis of an appeal to the public at large to be left to pursue their research as a fundamentally beneficent community of workers on behalf of civilisation. The rhetoric was by no means always successful, and here I am focussing really on only one side of the argument, but it is useful nonetheless to form an idea of the political justifications for knowledge-producing practices. That is Volume II. Volume III is directly connected, exploring the ways in which the processes of experiment and experience led to a form of expertise (not just a body of knowledge, but a way of coming to know) that was both exclusive and unassailable. The scientific answer to its critics often took the form of a claim to knowledge that, while not private, was not within the grasp of the layperson. If the morality, beneficence, humanity or utility of experimental processes of knowledge production could not be accessed by the non-scientific community, it did not mean that science was amoral or immoral, harmful, monstrous or useless. It was just that only another scientist was fit to judge the value of scientific work. This neat piece of sealed logic could seem dismissive and disingenuous, but it was often at the core of practitioners' scientism. At the same time, throughout the nineteenth century the scientist as both specialist and professional – that is to say, earning a living doing a particular disciplinary aspect of science – became more common and more respectable in society. Authority inhered in expertise, but also in the figure of the scientist himself (the gendering was almost always male). A certain temperament or mien, a certain facility of hand or eye, or a certain ability to process information, made the scientist into a leading figure of society, to rival the well-to-do and the clergy in pronouncing on the way of things in general. From matters of public health to common beliefs, the scientist increasingly took the role of expert commentator, but with an authority seated not in privilege or faith or tradition, but in the rigours of knowledge and its pursuit. In this, some saw the collapse of civilisation, while others saw its potential to flourish. Again, taking a broad view, the sources here permit a small insight into the ways in which scientists claimed political territory for themselves as social actors, on the basis, it must be repeated, of their curiosity, their uncertainty and above all their humanity, however misdirected it may have been.

Notes

1 The controversial (to some) starting points for this view might be thought to lie with Bruno Latour and Steve Woolgar, *Laboratory Life: The Construction of Scientific Facts* (Los Angeles: Sage, 1979); Bruno Latour, *Science in Action: How To Follow Scientists and Engineers Through Society* (Cambridge: Harvard University Press, 1987). The other classical pillar would be Thomas Kuhn, *The Structure of Scientific Revolutions* (Chicago: University of Chicago Press, 1962). See also, Sal Restivo and Julia Louhlin, 'The Invention of Science', *Cultural Dynamics*, 12 (2000): 135–49.

2 Karin Knorr Cetina, *Epistemic Cultures: How the Sciences Make Knowledge* (Cambdridge: Harvard, 1999); Ludwig Fleck, *Entstehung und Entwicklung einer wissenschaftlichen Tatsache: Einführung in Denkstil und Denkkollektiv* (Frankfurt: Suhrkamp, 1980); Lorraine Daston, 'The Moral Economy of Science', *Osiris*, 10 (1995): 2–24.

3 Steven Shapin, *The Scientific Life: A Moral History of A Late Modern Vocation* (Chicago: University of Chicago Press, 2008).

4 Jack Morrell and Arnold Thackray, *Gentlemen of Science: Early Years of the British Association for the Advancement of Science* (Oxford: Clarendon, 1981); Heather Ellis, *Masculinity and Science in Britain, 1831–1918* (London: Palgrave, 2017); Rob Boddice, 'Vivisecting Major: A Victorian Gentleman Scientist Defends Animal Experimentation, 1876–1885', *Isis*, 102 (2011): 215–37.

5 Stephen G. Bush, *The History of Modern Science: A Guide to the Second Scientific Revolution, 1800–1950* (Ames: Iowa State University Press, 1988); Enrico Bellone and Mirella Giacconi, *A World on Paper: Studies on the Second Scientific Revolution* (Cambridge: MIT Press, 1980).

6 Rob Boddice, *Edward Jenner* (Stroud: History Press, 2015); Rob Boddice, *The Science of Sympathy: Morality, Evolution, and Victorian Civilization* (Urbana Champaign: University of Illinois Press, 2016); Rob Boddice, *Humane Professions: The Defence of Experimental Medicine, 1876–1914* (Cambridge: Cambridge University Press, 2021).

7 Nadja Durbach, *Bodily Matters: The Anti-Vaccination Movement in England, 1853–1907* (Durham: Duke University Press, 2004); E.P. Hennock, 'Vaccination Policy Against Smallpox, 1835–1914: A Comparison of England with Prussia and Imperial Germany', *Social History of Medicine*, 11 (1998): 49–71; Naomi Williams, 'The implementation of compulsory health legislation: infant smallpox vaccination in England and Wales, 1840–1890', *Journal of Historical Geography*, 20 (1994): 396–412.

8 Joan Austoker and Linda Bryder, eds, *Historical Perspectives on the Role of the MRC: Essays in the History of the Medical Research Council of the United Kingdom and Its Predecessor, the Medical Research Committee* (Oxford: Oxford University Press, 1989); A. Landsborough Thomson, 'Origin of the British Legislative Provision for Medical Research', *Journal of Social Policy*, 2 (1973): 41–54.

9 Stefanos Geroulanos and Todd Meyers, *The Human Body in the Age of Catastrophe: Brittleness, Integration, Science, and the Great War* (Chicago: University of Chicago Press, 2018).

10 W.F. Bynum, *Science and the Practice of Medicine in the Nineteenth Century* (Cambridge: Cambridge University Press, 1994); Michael L. Geison, *Michael Foster and the Cambridge School of Physiology: The Scientific Enterprise in Late Victorian Society* (Princeton: Princeton University Press, 1978); Ruth Barton, '"Men of Science": Language, Identity and Professionalization in the Mid-Victorian Scientific Community', *History of Science*, 41 (2003): 73–119; J.C. Waller, 'Gentlemanly men of science: Sir Francis Galton and the professionalization of the British life-sciences', *Journal of the History of Biology*, 34 (2001): 83–114; J.B. Morrell, 'Professionalisation', *Companion to the History of Modern Science*, eds. G.N. Cantor, et al (London: Routledge, 1989); Heather Ellis, 'Knowledge, character and professionalisation in nineteenth-century British science', *Politics, Professionals and Practitioners*, eds. Wendy Robinson, Rob Freathy, Jonathan Doney (London: Routledge, 2018).

11 Rosemary Stevens, *Medical Practice in Modern England: The Impact of Specialization and State Medicine* (New Haven: Yale University Press, 1971); George Weisz, *Divide and Conquer: A Comparative History of Medical Specialization* (Oxford: Oxford University Press, 2006).

12 Peter J. Bowler, *Charles Darwin: The Man and His Influence* (Cambridge: Cambridge University Press, 1990); Peter J. Bowler, *The Eclipse of Darwin: Anti-Darwinian Evolution Theories in the Decades around 1900* (Baltimore: Johns Hopkins University Press, 1983); Gowan Dawson, *Darwin, Literature and Victorian Respectability* (Cambridge: Cambridge University Press, 2007); Alistair Sponsel, *Darwin's Evolving Identity: Adventure, Ambition, and the Sin of Speculation* (Chicago: University of Chicago Press, 2018).

13 Simon Szreter, 'The Importance of Social Intervention in Britain's Mortality Decline *c.* 1850–1914: A Re-interpretation of the Role of Public Health', *Social History of*

Medicine, 1 (1988): 1–38; R.J. Lambert, 'A Victorian National Health Service: State Vaccination 1855–71', *Historical Journal*, 5 (1962): 1–18.

14 Sandra Herbert, *Charles Darwin, Geologist* (Ithaca: Cornell University Press, 2005); Joe D. Burchfield, *Lord Kelvin and the Age of the Earth* (Chicago: University of Chicago Press, 1990); Howard Carlton, *Cosmology and the Scientific Self in the Nineteenth Century: Astronomic Emotions* (London: Palgrave, 2022).

15 Mark Neuendorf, *Emotions and the Making of Psychiatric Reform in Britain, c. 1770–1820* (London: Palgrave, 2021); Amelia Bonea, et al, *Anxious Times: Medicine and Modernity in Nineteenth-Century Britain* (Pittsburgh: University of Pittsburgh Press, 2019); Kurt Danziger, *Construction the Subject: Historical Origins of Psychological Research* (Cambridge: Cambridge University Press, 1994); Rick Rylance, *Victorian Psychology and British Culture, 1850–1880* (Oxford: Oxford University Press, 2000).

16 Steve Sturdy and Roger Cooter, 'Science, Scientific Management, and the Transformation of Medicine in Britain, *c.* 1870–1950', *History of Science*, 36 (1998): 421–66; Andrew Cunningham and Perry Williams, eds, *The Laboratory Revolution in Medicine* (Cambridge: Cambridge University Press, 1992); Sven Dierig, 'Engines for Experiment: Laboratory Revolution and Industrial Labor in the Nineteenth-Century City', *Osiris*, 18 (2003): 116–34.

17 Shane McCorristine, ed., *Spiritualism, Mesmerism and the Occult, 1800–1920* (New York: Routledge, 2016); Alison Winter, *Mesmerized: Powers of Mind in Victorian Britain* (Chicago: University of Chicago Press, 2000); John van Wyhe, *Phrenology and the Origins of Victorian Scientific Naturalism* (New York: Routledge, 2017); James Poskett, *Materials of the Mind: Phrenology, Race, and the Global History of Science, 1815–1920* (Chicago: University of Chicago Press, 2019); Hilary Grimes, *The Late Victorian Gothic: Mental Science, the Uncanny, and the Scenes of Writing* (London: Routledge, 2016); Tatiana Kontou, ed., *The Ashgate Research Companion to Nineteenth-Century Spiritualism and the Occult* (London: Routledge, 2016), esp. part 1.

18 R.P. Gaynes, *Germ Theory: Medical Pioneers in Infectious Diseases* (Washington D.C.: ASM Press, 2011); Margaret Pelling, 'Contagion/Germ Theory/Specificity', *Companion Encyclopedia of the History of Medicine*, eds. W.F. Bynum and Roy Porter (London: Routledge, 1993), I: 309–34; Michael Worboys, *Spreading Germs: Disease Theories and Medical Practice in Britain, 1865–1900* (Cambridge: Cambridge University Press, 2000);

19 Michael Worboys, 'Joseph Lister and the performance of antiseptic surgery', *Notes and Records of the Royal Society*, 67 (2013): 199–209; Stephanie Snow, *Blessed Days of Anaesthesia: How Anaesthetics Changed the World* (Oxford: Oxford University Press, 2008).

20 Quentin R. Skrabec, *The Metallurgic Age: The Victorian Flowering of Invention and Industrial Science* (Jefferson: McFarland & Co., 2006); Shane Ewen, 'Socio-technological disasters and engineering expertise in Victorian Britain: the Holmfirth and Sheffield floods of 1852 and 1864', *Journal of Historical Geography*, 46 (2014): 13–25; S. Arapostathis, 'Electrical innovations, authority and consulting expertise in late Victorian Britain', *Notes and Records of the Royal Society Journal of the History of Science*, 67 (2013): 59–76.

21 For a general account, see John V. Pickstone, *Ways of Knowing: A New History of Science, Technology, and Medicine* (Chicago: University of Chicago Press, 2001); Jutta Schickore, *About Method: Experimenters, Snake Venom, and the History of Writing Scientifically* (Chicago: University of Chicago Press, 2017).

22 L. Daston, 'British responses to psycho-physiology, 1860–1900', *Isis*, 69 (1978): 192–208; Judy Johns Schloegel and Henning Schmidgen, 'General Physiology, Experimental Physiology, and Evolutionism: Unicellular Organisms as Objects of Psychophysiological Research, 1877–1918', *Isis*, 93 (2002): 614–45; Merriley Borell, 'Instrumentation and the Rise of Modern Physiology', *Science & Technology Studies*, 5 (1987): 53–62.

23 William J. Ashworth, 'The calculating eye: Baily, Herschel, Babbage and the business of astronomy', *The British Journal for the History of Science*, 27 (1994): 409–41; Carlton, *Cosmology*; Simon Schaffer, 'Herschel in Bedlam: Natural History and Stellar Astronomy', *The British Journal for the History of Science*, 13 (1980): 211–39.

24 S.M. Stigler, *The History of Statistics: The Measurement of Uncertainty before 1900* (Cambridge: Harvard University Press, 1986); Theordore M. Porter, *Karl Pearson: The Scientific Life in a Statistical Age* (Princeton: Princeton University Press, 2004).

25 Douglas A. Lorimer, *Science, race relations and resistance: Britain, 1870–1914* (Manchester: Manchester University Press, 2013); Debbie Challis, *The Archaeology of Race: The Eugenic Ideas of Francis Galton* (London: Bloosmbury, 2013); G.R. Searle, *Eugenics and Politics in Britain, 1900–1914* (Leyden: Noordhoff International, 1976); Pauline M.H. Mazumdar, *Eugenics, Human Genetics and Human Failings: The Eugenics Society, Its Sources and Its Critics in Britain* (London: Routledge, 1992); Robert A. Nye, 'The Rise and Fall of the Eugenics Empire: Recent Perspectives on the Impact of Biomedical Thought in Modern Society', *Historical Journal*, 36 (1993): 687–700.

26 Evelleen Richards, *Darwin and the Making of Sexual Selection* (Chicago: University of Chicago Press, 2017).

27 Jan Golinski, *Science as Public Culture: Chemistry and Enlightenment in Britain, 1760–1820* (New York: Cambridge University Press, 1992); Colin A. Russell, ed., *Chemistry, Society and Environment: A New History of the British Chemical Industry* (Cambridge: Royal Society of Chemistry, 2000).

28 Peter M. Harman, *Energy, Force and Matter: The Conceptual Development of Nineteenth-century Physics* (Cambridge: Cambridge University Press, 1982); Robert D. Purrington, *Physics in the Nineteenth Century* (New Brunswick: Rutgers University Press, 1997).

29 Heather Ellis and Ulrike Kirchberger, eds, *Anglo-German Scholarly Networks in the Long Nineteenth Century* (Leiden: Brill, 2014); Brett M. Bennett and Joseph M. Hodge, eds, *Science and Empire: Knowledge and Networks of Science across the British Empire, 1800–1970* (Houndmills: Palgrave, 2011); Ruth Schilling and Kay Peter Jankrift, 'Medical Practice in Context: Religion, Family, Politics and Scientific Networks', *Medical Practice, 1600–1900: Physicians and Their Patients*, ed. Martin Dinges, et al (Leiden: Brill, 2016): 131–48; Bernard Lightman and Gowan Dawson, eds, *Victorian Scientific Naturalism: Community, Identity, Continuity* (Chicago: University of Chicago Press, 2014); Robert J. Richards, *Darwin and the Emergence of Evolutionary Theories of Mind and Behaviour* (Chicago: University of Chicago Press, 1987); Luis A. Campos, *Radium and the Secret of Life* (Chicago: University of Chicago Press, 2015).

30 Adrian Desmond and James Moore, *Darwin: The Life of a Tortured Evolutionist* (New York: Norton, 1994); Porter, *Karl Pearson*; Nicholas Wright Gillham, *A Life of Sir Francis Galton: From African Exploration to the Birth of Eugenics* (Oxford: Oxford University Press, 2001); Geoffrey Cantor, *Michael Faraday: Sandemanian and Scientist* (Houndmills: Palgrave, 1993); Taliah Drayak, *Mary Somerville: The Queen of Nineteenth Century Science* (Barnsley: Pen & Sword, 2019).

31 Elsbeth Heaman, *St Mary's: The History of a London Teaching Hospital* (Montreal: McGill-Queen's University Press, 2003); Geison, *Michael Foster and the Cambridge School of Physiology*; Harriette Chick, Margaret Hume, Marjorie MacFarlan, *War on Disease: A History of the Lister Institute* (London: A. Deutsch, 1971); Camilla Mork Rostvik, 'The changing power of scientific institutions: The modern histories of *Nature* and The Royal Society', *Studies in History and Philosophy of Science*, 61 (2017): 46–9; Adrian Tinniswood, *The Royal Society and the Invention of Modern Science* (New York: Basic Books, 2019); Ray Desmond, *The History of the Royal Botanic Gardens Kew* (London: Royal Botanic Gardens, 2007).

32 Frederic L. Holmes and Trevor Harvey Levere, eds, *Instruments and Experimentation in the History of Chemistry* (Cambridge: MIT Press, 2000); *Scientific Instruments and Collections*, general ed, A.D. Morrison-Low, 5 vols (Leiden: Brill, 2009–21).
33 Erika Lorraine Milam and Robert A. Nye, eds, 'Scientific Masculinities', *Osiris*, 30 (2015).
34 Weisz, *Divide and Conquer*; A. Tuchman, *Science, medicine and the state in Germany: The case of Baden, 1815–1871* (Oxford: Oxford University Press, 1993); Thomas D. Brock, *Robert Koch: A Life in Medicine and Bacteriology* (Madison: Science Tech Publishers, 1988); Gabriel Finkelstein, *Emil Du Bois-Reymond: Neuroscience, Self, and Society in Nineteenth-Century Germany* (Cambridge: MIT Press, 2013); Frederic Lawrence Holmes, *Claude Bernard and Animal Chemistry: The Emergence of A Scientist* (Cambridge: Harvard University Press, 1974); Gerald Geison, *The Private Science of Louis Pasteur* (Princeton: Princeton University Press, 1995).
35 Peter Hobbins, *Venomous Encounters: Snakes, Vivisection and Scientific Medicine in Colonial Australia* (Manchester: Manchester University Press, 2017); Roy MacLeod, ed., 'Nature and Empire: Science and the Colonial Enterprise', *Osiris*, 15 (2000).
36 The phrase is from Daston and Galison, *Objectivity*; Paul White, 'Darwin's Emotions: The Scientific Self and the Sentiment of Objectivity', *Isis*, 100 (2009): 811–26.

INTRODUCTION TO VOLUME I
Curiosity

I begin toward the end, in France. These are questionable choices for a study of the long nineteenth century focusing on sources in English, but in this age great scholars and thinkers were able to reflect on an explosion of scientific endeavour that had paid little heed to national boundaries. At the end of his life, Ernest Renan, the giant of French Orientalism and scholar and commentator on seemingly every aspect of civilisation, dug up a manuscript that he had buried after the turmoil of 1848 and published it. The book appeared in 1891 and was immediately translated into English, entitled *L'Avenir de la Science, Pensées de 1848* (*The Future of Science, Thoughts of 1848*). It is thus a text from both the late and the middle period of my focus. Insofar as it reflected on intellectual life from 45 years prior to its publication, its appearance seemed to underscore the accuracy of its forecasts. The future of science in the 1840s was the present of science in the 1890s.

The seventh chapter of *The Future of Science* serves as the keynote here, for it both condemns and celebrates curiosity, both as a practice and as an element of scientific character. Renan does not exactly settle the moral aporia that surrounded curiosity, but he did effectively capture both the spirit of its contemporary value and the moral abyss that could be swept in under its direction. First, the criticism:

> If the many laborious workers to whom science owes its progress had possessed the philosophical spirit of what they did, if they had perceived in learning something more than the satisfaction of their vanity or curiosity, how many precious moments would have been saved, how many fruitless excursions spared, how many lives given up to insignificant works would have been devoted to more useful researches.

Curiosity, unbounded and undisciplined, unattached to a higher purpose, was, at best, a narcissistic practice. At worst, it was actively harmful, wasting time, wasting lives, wasting purpose. It risked the overthrow of morality, the furious pursuit of experiment for its own sake leaving all ethical and social considerations in the dust.

The mere 'collector' or the 'amateur' *might* be useful to science, but insofar as they pursue their science for their own satisfaction or pleasure, so they fall short of what science should be, victims of their own curiosity. Written at the high age

DOI: 10.4324/9781003009337-2

of the amateur, when science was pursued principally by independently wealthy gentlemen generalists, who were as likely to designate themselves philosophers as scientists, this was a pointed charge. Renan, though, was aware that there were 'various degrees of curiosity'. When it leaves behind the 'paltry instinct of the collector' and becomes 'a love of knowledge' it can be the making of a 'very noble existence'. Whatever personal pleasure was wrapped around this kind of curiosity could drive the production of useful new knowledge.

> How many compilations, precious in view of ultimate researches would have never been made but for that innocent love of labour with which many mildly active natures cheat their craving. It would be cruel indeed to refuse to those humble workers the trivial and not very lofty pleasure.

Here lay the kernel of *satisfaction* that, for all of the drudgery of scientific labour, provided its ethic. A keynote of scientific research in the nineteenth century is tedium, endurable only by practitioners of a particular character who could find, within this tedium, a singular, irresistible attraction. Yet, in Renan's view, the value of science could not lie in this attraction, this curiosity, itself, for the satisfaction of the scientist is ultimately immaterial to the status of the knowledge produced. Curiosity is not, a priori, *moral*. To 'take up science as merely interesting and curious is to humiliate it'. A higher purpose is intrinsic to the scientific pursuit 'to know and to love that which is'. Curiosity has to transcend self-interest if it is to have value.

It is with these passages from Renan's *Future of Science* that this volume begins. There are, throughout, a variety of expressions about curiosity that fall on one side or another of Renan's fine line. Into the mix one must add the context of materialism and atheism. Renan's ideal scientist, in quest of perfection, ultimately did so in service of God. In increasing numbers, scientists in the later decades of the nineteenth century pursued knowledge precisely on the assumption that there was no God; that nature could, in the fullness of time, simply be revealed by stolid, determined work, buoyed on the indefatigable energy supplied by scientific curiosity.

The materialist aspect more acutely focuses the status of scientific knowledge production on the figure of the scientist and on scientific equipment: the means of neutrality, as it were. Individual examples of scientific virtue and vice (often at the same time, depending on the point of view) dot the nineteenth century, from Claude Bernard to Charles Darwin, but perhaps more significant is the emergence of the scientific self as a trope and of detailed accounts of new scientific equipment that was supposed, in effect, to enable curiosity while removing prejudice. Both scientists and their anxious opponents could refer to the man of science in immediately identifiable terms, as a diligent rational servant of a noble curiosity, for the sake of society in general, or as an amoral, unrestrained monster, pursuing new facts and rehearsing old ones for kicks, lost, it would seem, in his machines. From the point of view of a rapidly professionalising scientific academy, publicly and privately funded, the ends justified the means. Indeed, the means, however ethically dubious, could be placed into a calculus of suffering. New knowledge would, assuredly, lessen the amount of suffering in the world, and could therefore justify a certain amount of suffering in the pursuit of it. And the moral qualities of the scientist would assure that this certain amount, indefinable and ambiguous, would not be extended indefinitely into clear practices of monstrousness, such as human experimentation or the extermination of the 'weak'. From the point of view of those who viewed the growth of scientific experimentation through critical and often religious eyes, science had already and long-since over-extended what was morally justifiable in pursuit of its ends, which could not ever justify these means. At the beginning of the period, there were still active debates about the sanctity of the dead body and the scientific employment of human remains for experimental purposes. The materialist could find no moral compunction in subjecting the executed criminal body to anatomical, electrical and chemical investigation, but to the religious upholder of a certain moral view of civilisation, this very lack of compunction, combined with the desecration of the dead, was an alarm signal about a present and growing danger. Whither morality? As physiology, toxicology, immunology and bacteriology all took off in the second half of the nineteenth century, shifting the experimental subject from the dead human to the living animal, the prevailing fears appeared fully justified. Science had lost its compassion, and with it its moral compass and its capacity to see and understand suffering, while blindly groping in its curiosity for suffering's cure and for answers from nature.

The scientific rationale, across many different specialisms, did not overly focus on the outmoded cosmology of prevailing Judeo-Christianity and its understanding of 'man's place in nature', but rather emphasised an overwhelming feeling of the possibility of scientific revelation. Society, civilisation, social policy from public health to sexual relations and childbirth, to industrial regulation and geopolitics could be run anew on the principles of empiricism. Scientific controls and the extraordinary reasoning powers of its practitioners had shown and would continue to show that public 'faith' was much better directed in its leadership that it was surrendered up to the immaterial unknown machinations of an unprovable deity. The evolutionary idiom, in particular, often veiled its criticism of theism in favour of documented revealed facts. The upshot was that while humanity remained, per the Christian 'great chain of being', atop the other animals in terms of evolutionary advancement, all notion of beings above the human disappeared. The existence of angelic and divine entities, while not given up by every scientist by any means, was nonetheless beyond the purview of what could be investigated. No amount of curiosity could lead to discovery and, as such, they became irrelevances in the quest for new knowledge. The agnostic declaration was the abandonment of supernatural curiosity. Logically, the divine became irrelevant for how society should function and how it should be led. Insofar as the scientific community saw its curiosity as a practice of beneficence, the rewards of which were to be reached at some indefinite future point, they implicitly worked against the principles of social and political cohesion that was rooted in Christian morals and teaching.

This is not to re-establish a polarity of science and religion that has long-since been rightly complicated.[1] Science and religion were constantly and messily entangled throughout the nineteenth century. It is simply to say that religious critics of science in particular understood there to be a specific risk in an unbridled and entirely earthly curiosity, from the point of view not only of the erosion of faith, but from the point of view of social and political leadership. What could there be, at the core of materialism, around which to build a social fabric or shared sense of morality or social contract? The scientifically curious tried to address this question with specific claims that to better understand nature, the world, the stars, would permit a better understanding of how society functions. How much better to lead the world from the point of view of a real, proven understanding, than to lead it from the point of view of theological speculation?

The reality, of course, was of competing institutionalised hubris. Declarations and decisions were regularly made on the back of faulty science in faith that the science was unimpeachable. At such moments, the social, cultural and political history of science emerges clearly, for it becomes more a question of what is done in the name of science than of any notion of progress, innovation or improvement. The knowledge that scientists produce claims the status of knowledge, often inflexibly, even when that knowledge turns out to be partial, faulty or plain wrong. And in the meantime, the conviction that it is impartial and correct undergirds opinions, policies, social tensions and debates, and further fuels more curiosity in the same vein.

These are generalist and impressionistic lines. No one source captures them completely. But immersion in these kinds of sources (of which these in this volume are but a representative example) speaks to their truth. If there is a takeaway here, it concerns an appreciation of what was at stake with the pursuit of scientific curiosity in the experimental idiom. It is not merely that the scientific community was itself shifting in terms of its specialisation, professionalisation and internationalisation, though these things are intrinsically important to an understanding of the history of science. More important is the degree to which these processes were enmeshed in much bigger conversations and conflicts about what humanity was, what its status was and who had the moral authority to lead it, shape it, guide it and instruct it. The scientific ability to see farther in space and farther in time, to see grander things and microscopic things, and to manufacture knowledge through experimental design pushed scientists to innovate and take risks. The sanctity of the living body and the divine nature of the cosmos were both subject to material scrutiny, such that the sheer fact of investigation was, to some, a moral outrage, irrespective of the results. Curiosity, then, was a process, ever being refined, that equated to the promise of a better society – better understood and better led. To those endowed with such a quality and provided with the necessary means, it was an obvious virtue: a central quality of the scientific self. To those on the outside, the fabric of society and civilisation seemed at risk.

Note

1 For a recently updated overview, see Thomas Dixon and Adam Shapiro, *Science and Religion: A Very Short Introduction* (Oxford: Oxford University Press, 2022).

Part 1

GENERAL

Editorial Headnote

Renan, Ernest, *The Future of Science: Ideas of 1848* (London: Chapman and Hall, 1891)

1

THE FUTURE OF SCIENCE
Ideas of 1848

Ernest Renan

Source: London: Chapman and Hall, 1891, 109–116

JUST as in the very bosom of religion there are a great many men handling sacred matters without the least idea of their lofty meaning, and only looking at them in the light of vulgar manipulation, so in the field of science there are labourers—very worthy people in the main—often utterly lacking in the sentiment of their work and its value in the furtherance of the ideal. Let us hasten to add that it would be unfair to require of the *savant* the ever immediate consciousness of the aim of his work; it would be bad taste on our part to wish him to speak of it deliberately in and out of season; it would be compelling him to head all his works with the self-same prolegomena. Take the most beautiful scientific works, peruse the works of Letronne, of Burnouf, of Lassen, of Grimm and of the princes of modern criticism generally and perhaps you would seek in vain for a page *directly* and *abstractedly* philosophical. The fact of their authors being thoroughly imbued with the philosophic spirit is made manifest not so much by an isolated flight of rhetoric as by the general spirit and method. Nay, this prudent abstention may often be taken as an act of scientific virtue and the heroes of science are they who, though more capable than no matter whom of indulging in lofty speculation, have the strength of mind to confine themselves to the strict statement of facts and abstaining voluntarily from anticipated generalizations. Works undertaken without this noble spirit may even powerfully aid in the cultivation of the human intellect irrespective of the more or less paltry intentions of their authors. Is it at all necessary for the marble quarrier to have an idea of the future monument of which the blocks which he quarries will form a part? Among the laborious workers of science who have raised the edifice of science many looked no further than the stone they shaped, or at any rate no further than the limited region where they placed it. Like the ants, they each bring their individual tribute, remove some obstacle, cross one another incessantly, apparently without the slightest attempt at order, and only get in one another's way. Nevertheless, it happens that by the united labour of so many men and without any preconceived plan a science is organized, and organized in noble proportions. An invisible genius has been the architect presiding over the whole and has made all those isolated efforts subservient to a perfect unity.

DOI: 10.4324/9781003009337-4

By studying the origin of each science one will find that the first steps were nearly always taken with no very distinct consciousness, and that among others, philological studies owe a large debt of gratitude to very mediocre intellects, which at the outset laid down their material conditions. Hervas, Paulin de Saint-Barthélemi, Pigafetta who must be considered the founders of linguistic science were certainly not geniuses. What an immense fact in the history of the human intellect is the initiation of the Latin world in the understanding of Greek literature. The two men who contributed most powerfully to it, Barlaam and Léontius Pilatus were according to Petrarch and Boccacio who knew them so very intimately two nonentities who were as surly as they were whimsical. The majority of the Greek emigrants who played so important a part in the development of European intelligence were men of no parts whatsoever, downright labourers who took advantage *per alcuni denari* of their knowledge of Greek. For every Bessarion there were a hundred Philelphuses. The lexicographers are as a rule not very great philosophers and still the best book on generalities has not had so great an influence on the higher sciences as the dictionary – philosophically an exceedingly poor performance – by which Wilson made the study of Sanskrit possible in Europe. There are works requiring an amount of plodding drudgery to which men impelled by too powerful a craving for philosophical studies would with difficulty submit. Would lofty and energetic minds have succeeded in accomplishing the immense works issued from the scientific workshops of the congregation of St. Maur?[1] Every scientific work, conducted according to a sound method, keeps its unquestionable value, irrespective of the wider or narrower views of its author. The only useless works are those in which the smatterer or the quack pretends to imitate the bearing of genuine science, and those in which the author, prompted by an interested thought, or by the preconceived dreams of his imagination is bent at all costs upon finding his chimeras everywhere.

Though it is not necessary for the workman to have a perfect knowledge of the work he executes, one could wish those who devote themselves to special labours to have an idea of the whole which alone imparts value to their researches. If the many laborious workers to whom science owes its progress had possessed the philosophical spirit of what they did, if they had perceived in learning something more than the satisfaction of their vanity or curiosity, how many precious moments would have been saved, how many fruitless excursions spared, how many lives given up to insignificant works would have been devoted to more useful researches. When we come to consider that the intellectual labour of whole centuries, of whole countries, of Spain for instance, has been wasted, for lack of a substantial object in view, that millions of volumes have crumbled to dust without producing the least result, we cannot help regretting deeply this immense loss of human force, brought about by the absence of guidance and for want of a distinct perception of the goal to be reached. The intensely sad impression one feels on entering a library is due mainly to the thought that nine-tenths of the books crowded together there have missed their mark, and whether through the fault of the author himself, or through the fault of circumstances have never had nor will ever have, any direct action on the onward march of humanity.

It seems to me that science will only then recover its dignity when it takes up its definite stand on the grand and wide point of view of its veritable aim. In former days there was room for that trivial and more or less innocuous character of the *savant* of the Restoration, a semi-courtier-like part, the actor of which had a way of allowing himself to be taken for a man of *solid* attainments who tossed his head at ambitious innovations; it was a way of securing the patronage of some Mæcenas, a duke or peer, who as a mark of high favour admitted him in the capacity of a bit of furniture of his drawing-room, or of an antique *curio* of his collection; and in the whole transaction there was nothing deserving the name of serious, there was nothing but the more or less imbecile laughter of vanity, the laughter so terribly aggravating when it presumes to meddle with serious matters. That is the kind of thing which is doomed to disappear for ever more, that is as good as buried together with the playthings of a society in which shams still played so important a part. To entice science from the grandiose midst of humanity in order to make it an idle plaything of a court or a drawing-room is degrading science, for the day is not far distant when everything that is not serious and true will be ridiculous. Let us, then, be true for Heaven's sake, true like Thales when from his own initiative and impelled from within he took to speculating upon nature, true like Socrates, true like Jesus, genuine like St. Paul, genuine like all those great men which the ideal claimed as its own, took possession of and dragged after it. Let us leave old-world twaddle to plead lukewarmly in apology for science that like everything else, it is a necessity, an ornament, that it confers lustre on a country, etc., etc. All this is so much sheer imbecility. Where is the noble and philosophic soul, eager for perfection, possessing the sentiment of its intrinsic worth that would consent to sacrifice itself to such vanities, that would gladly take up its place in the inanimate tapestry of human nature, to enact in the living world the part of a mummy in a museum. As for myself, I confess frankly that if I could see a form of life more beautiful than that of science, I should run to embrace it. How can one be reconciled to what one knows to be the second-best? How can one consent to put one's self among the waste, to accept a mere part of show when life is so short, when nothing can make up for the moments not devoted to the delights of the ideal? Oh, truth and earnestness of life, oh holy poesy of things, what is there to console us for having no feeling for thee? And to talk for a moment of the serious hour to which everyone must look forward to appreciate things in their true light, who will be able to breathe his last tranquilly, when, in casting a glance backward, he only finds in his life frivolity or gratified curiosity? The end only is worthy of being looked at, the whole rest nothing but vanity. To live does not mean to glide smoothly along a pleasant surface, it does not mean making the world your plaything in the pursuit of your own pleasure; it means to partake of many beautiful things, it means to be the fellow-traveller of the stars; it means knowledge, hope, love, admiration;

it means well-doing. He has lived most and most worthily who by his heart and intellect, by his acts has worshipped most.

Hence, to look at science in the light of a mere satisfaction of vanity or curiosity is as great an error as to consider poetry merely the listless exercise of frivolous minds, or literature the amusement of which one tires least, and to which one comes back most willingly. The collector and the amateur may render signal services to science, but they are neither *savants* nor philosophers. They are as far removed from it as the manufacturer. For they amuse themselves, they pursue their own pleasure just as the manufacturer seeks his profit. I am aware of there being various degrees of curiosity. There is a wide difference between the paltry instinct of the collector which scarcely differs from the attachment of the child to his toys, and the more elevated form when it becomes a love of knowledge, that is; a legitimate instinct of nature and may be productive of a very noble existence. Boyle and Charles Nodier are only mere inquirers, and still they closely trench upon the philosopher. It is very rare, in fact, that with the highest exercise of the intellect there is not mixed up a certain amount of pleasure, which though having no value as far as the ideal is concerned, is none the less useful. It would be difficult to say how many discoveries have been simply due to curiosity. How many compilations, precious in view of ultimate researches would have never been made but for that innocent love of labour with which many mildly active natures cheat their craving. It would be cruel indeed to refuse to those humble workers the trivial and not very lofty pleasure, but sweet withal, which M. Daunou has so well defined as *paperasser*.[2] We are all more or less glad at having felt the same satisfaction, if for no other reason than because it has helped us to devour the barren pages of science. Without this the first studies necessary to master the material baggage of a language would be unbearable, and thanks to this they become the most attractive imaginable.

We may go further still and positively affirm that without this attraction the most learned men of modern times who were neither borne up by lofty philosophical views, nor by directly religious motives, would not have undertaken those immense labours which have made the investigations of higher criticism possible to us. He who, with our material wants intensified to a greater pitch, would accomplish a similar act of abnegation to-day, would be a hero. But it is important to maintain for all that, that this curiosity has no immediate moral whatsoever, and that its possession does not constitute the *savant*. There are manufacturers who exploit science for their benefit, the others exploit it for their amusement. No doubt the latter is the better part, but there is not an enormous difference between the two. Pleasure being essentially personal and interested, there is nothing whatsoever sacred or moral about it. All literature, all poesy, all science whose only aim is to amuse or merely to awaken interest is for this very reason frivolous and vain, or to speak correctly, has no right at all to call itself literature, poesy or science. The mountebanks in the streets and elsewhere do as much, and what is more, succeed better. How is it that one considers the perusal of Corneille, Goethe, Byron as a serious occupation, and that the perusal of such and such a modern novel or drama is looked upon as merely a pastime? For the same reason that the *Quarterly* and *Edinburgh Reviews* are serious periodicals and that the *Magasin Pittoresque* is a frivolous book.

Therefore, to take up science as merely interesting and curious is to humiliate it. In that case Christian asceticism would be perfectly right in its opposition. The sole legitimate means of making one's self the apologist of science is to look upon it as the essential element to human perfection. "The Imitation of Christ," after having begun like the "master of those who know" with the words; "Every man is naturally desirous to know," was thoroughly right in adding; "But what is the use of science without love? Better by far the humble peasant who serves God, than the proud philosopher who watches the course of the stars and neglects himself. What is the use of the knowledge of things, the ignorance of which will not cause us to be condemned? All is vanity, save the love of God and to serve Him." There is no gainsaying this, if science be conceived as a simple series of formulas, if perfect love be possible without knowledge. If we place perfection on one side, on the other vanity, how can one fail to follow perfection? But it is this very division which is illegitimate, because perfection is impossible without science. The true way of worshipping God, is to know and to love that which is.

Notes

1 The congregation of St. Maur, a reformed branch of the order of St. Benedict, was founded in France in the seventeenth century and has rightly been called the nursery ground of *savants*. – Transl.

2 *Paperasser* means both a love of scribbling and of ferreting among documents. Daunou was the author of a remarkable series of Historical Essays; died about 1840. – Transl.

Editorial Headnote

George Craik, *The Pursuit of Knowledge Under Difficulties* (London: Charles Knight, 1831), II, 1–3, 121–124

George Lillie Craik (1798–1866), Scottish literary critic, noted for his books on Edmund Spenser and Francis Bacon.

2

THE PURSUIT OF KNOWLEDGE UNDER DIFFICULTIES

George Craik

Source: London: Charles Knight, 1831, II, 1–3, 121–124

WE remarked, at the close of our former volume, that the moral habits which the Pursuit of Knowledge has a tendency to create and foster, form one of its chief recommendations. Knowledge is, essentially and directly, power; but it is also, indirectly, virtue. And this it is in two ways. It can hardly be acquired without the exertion of several moral qualities of high value; and, having been acquired, it nurtures tastes, and supplies sources of enjoyment, admirably adapted to withdraw the mind from unprofitable and corrupting pleasures. Some distinguished scholars, no doubt, have been bad men; but we do not know how much worse they might have been, but for their love of learning, which, to the extent it did operate upon their characters, could not have been otherwise than beneficial. A genuine relish for intellectual enjoyments is naturally as inconsistent with a devotion to the coarser gratifications of sense, as the habit of assiduous study is with that dissipation of time, of thought, and of faculty, which a life of vicious pleasure implies.

But knowledge is also happiness, as well as power and virtue; happiness both in the acquisition and in the possession. And were the pursuit of it nothing better than a mere amusement, it would deserve the preference over all other amusements, on many accounts. Of these, indeed, the chief is, that it must almost of necessity become something better than an amusement,—must invigorate the mind as well as entertain it, and refine and elevate the character while it gives to listlessness and weariness their most agreeable excitement and relaxation. But, omitting this consideration, it is still of all amusements the best, for other reasons. So far from losing any part of its zest with time, the longer it is known the better is it loved. There is no other pastime that can be compared with it in variety. Even to him who has been longest conversant with it, it has still as much novelty to offer as at first. It may be resorted to by all, in all circumstances; by both sexes, by the young and the old, in town or in the country, by him who has only his stolen half hour to give to it, and by him who can allow it nearly his whole day, in company with others, or in solitude, which it converts into the most delightful society. Above all, it is the cheapest of all amusements, and consequently the most universally accessible.

DOI: 10.4324/9781003009337-5

Causes which will suggest themselves to the reflection of every reader, and which, therefore, we need not here stop to explain, have hitherto, in a great measure, excluded our labouring population from the enjoyments of science and literature; but this state of things is passing away, and the habit of reading is extending itself rapidly, even among the humblest ranks. Nothing can be more natural than this. A book is emphatically the poor man's luxury; for it is of all luxuries that which can be obtained at the least cost. By means of itinerating libraries for the country, and stationary collections for each of our larger towns, almost every individual of the population might be enabled to secure access for himself to an inexhaustible store of intellectual amusement and instruction, at an expense which even the poorest would scarcely feel. As yet these advantages have been chiefly in the possession of the middle classes, to whom they have been a source not more of enjoyment than of intelligence and influence.

Among the highest orders of society, the very cheapness of literary pleasures has probably had the effect of making them to be less in fashion than others of which wealth can command a more exclusive enjoyment. Even such distinction as eminence in intellectual pursuits can confer must be shared with many of obscure birth and low station; and on that account alone has doubtless seemed often the less worthy of ambition to those who were already raised above the crowd by the accidents of fortune. Yet, whatever enjoyment there may really be in such pursuits will not, of course, be the less to any one, because he happens to be a person of wealth or rank. On the contrary, these advantages are perhaps on no other account more valuable, than for the power which they give their possessor of prosecuting the work of mental cultivation to a greater extent than others. He has, if he chooses, a degree of leisure and freedom from interruption, greatly exceeding what the generality of men enjoy. Others have seldom more than the mere fragments of the day to give to study, after the bulk of it has been consumed in procuring merely the bread that perisheth; he may make literature and philosophy the vocation of his life. . . .

. . . Hence the creation of a new science, now commonly known by the name of Electro-Chemistry, being that which regards the supposed action of electricity in the production of chemical changes. The discourse in which these discoveries were unfolded was crowned by the French Institute with their first prize, by a decision which reflects immortal honour upon that illustrious body; who thus forgot not only all feelings of national jealousy, but even the peculiar and extraordinary hostility produced by the war which then raged between the two countries, in their admiration of genius and their zeal for the interests of philosophy.

But the results which this great chemist had already obtained only formed, in his hands, the source of new discoveries. In the interesting and extraordinary nature of its announcements, the Bakerian lecture of 1807 was as splendid a production as that of the former year. There are certain substances, as the reader is aware, known in chemistry by the name of alkalis, of which potash and soda are the principal. These substances, chemists had hitherto in vain exhausted their ingenuity, and the resources of their art, in endeavouring to decompose. The only substance possessing alkaline properties, the composition of which had been ascertained, was ammonia, which is a gas, and is therefore called volatile alkali; and this having been found to be a compound of certain proportions of hydrogen and nitrogen, an opinion generally prevailed that hydrogen would be found to be also a chief ingredient of the *fixed* alkalis. Davy determined, if possible, to ascertain this point, and engaged in the investigation with great hopes of success, from the surpassing powers of decomposition which he had found to belong to his new agent, the galvanic influence. The manner in which he pursued this object is one of the most beautiful specimens of scientific investigation on record. One of the most important of the laws of galvanic decomposition, which he had previously discovered, was, that, when any substance was subjected to this species of action, its oxygen (an ingredient which nearly all substances contain) was developed at what is called the positive end or pole of the current of electricity, while, whenever any hydrogen or inflammable matter was present, it uniformly appeared at the opposite or negative pole. Proceeding upon this principle, therefore, Davy set to work with a fixed alkali; and at first submitted it dissolved in water to the galvanic action. The result, however, was, that the water alone was decomposed, nothing being disengaged by the experiment but oxygen and hydrogen, the ingredients of that fluid, which passed off as usual, the former at the positive, the latter at the negative pole. In his subsequent experiments, therefore, Davy proceeded without water, employing potash in a state of fusion; and, having guarded the process from every other disturbing cause that presented itself, by a variety of ingenious arrangements, he had at last the satisfaction of seeing the oxygen gas developed, as before, at the positively electrified surface of the alkali, while at the same time, on the other side, small globules of matter were disengaged, having all the appearances of a metal. The long agitated question was now determined; the base of the fixed alkalis was clearly metallic. To ascertain the qualities of the metallic residue which he had thus obtained from the potash, was Davy's next object. From its great attraction for oxygen, it almost immediately, when exposed to the atmosphere, became an alkali again, by uniting with that ingredient; and at first it seemed on this account hardly possible to obtain a sufficient quantity of it for examination. But at last Davy thought of pouring over it a thin coating of the mineral fluid called naphtha, which both preserved it from communication with the air, and, being transparent, allowed it to be examined.

We have thus rapidly sketched the course of these brilliant and successful experiments, because they form a most interesting and instructive exemplification of the manner in which knowledge is pursued, and the secrets of nature extorted from her by well-directed interrogation. The business of philosophic experiment, it may be well to observe, is not a mere random expenditure of tests and applications. The true disciple of the inductive philosophy, on the contrary, has always in his contemplation, while conducting his experiments, an idea or end which he aims at realizing, and which, in fact, directs him to every experiment to which he resorts. Thus, in the present instance, the idea in Davy's mind was, that the alkali was compounded of two ingredients which had severally an attraction for the two opposite poles of the electric current. This idea he never lost sight of throughout the whole course of his experiments, though he repeatedly shifted his ground in regard to the contrivances by which he sought its proof and manifestation. To proceed in any other way would not be to philosophize, but merely, as it were, to dip the hand into the bag of chance in quest of a discovery, as men draw prizes at a lottery. It is true that, until the experiment has confirmed or refuted his expectations, this guiding idea upon which the experimenter proceeds must be regarded merely as a conjecture. But such a conjecture or hypothesis he must have in his mind, or he is in no condition to set about the inquisition of nature. What progress would the conductor of a trial in a court of justice be likely to make, in questioning a witness, without some previous notion of the truth which the evidence was likely to establish? He might waste the whole day in putting questions and receiving answers, and at last have ascertained nothing. Just as unprofitably would the interrogator of nature spend his time, if he had no directing anticipation in every case, according to which to order his experiments. Accident might, it is possible, throw a discovery in his way; but his own occupation would be evidently as idle and as little that of a philosopher as the rattling of a dice-box. *Whenever, indeed, a discovery is made without being anticipated, we say that it has been made by chance.* On the other hand, the history of all discoveries that have been arrived at by what can with any propriety be called philosophical investigation and induction, attests that necessity which has been asserted of the experimenter proceeding in the institution and management of his experiments upon a previous idea of the truth to be evolved. This previous idea is what is properly called a *hypothesis*, which means something *placed under* as a foundation or platform on which to institute and carry on the process of investigation. A *theory* is a completed view of a harmonious system of truths, evolved and proved by calculation or induction. As the latter is the necessary completion of every philosophical inquiry, so the former is its equally indispensable beginning. It is the aim in the mind of the philosopher, without which he cannot philosophize. It makes, in short, the main difference between the experiments of the philosopher in his laboratory, and those of the child among his play-things. Of course, however, every hypothesis must give way before an experiment the result of which cannot be reconciled with it.

Editorial Headnote

John Frederick William Herschel, *Preliminary Discourse on the Study of Natural Philosophy* (Philadelphia: Carey and Lea, 1831), 57–64

Sir John Frederick William Herschel, 1792–1871, was a giant of nineteenth-century science, noted for mathematics, astronomy and chemistry among other pursuits. This work was a keynote in the inductive method, which both foregrounded the importance of observation as well as keeping an open end in scientific conclusion. It might be thought the foundation of a sustainable scientific curiosity.

3

PRELIMINARY DISCOURSE ON THE STUDY OF NATURAL PHILOSOPHY

John Frederick William Herschel

Source: Philadelphia: Carey and Lea, 1831, 57–64

OF EXPERIENCE AS THE SOURCE OF OUR KNOWLEDGE.—OF THE DISMISSAL OF PREJUDICES.—OF THE EVIDENCE OF OUR SENSES

INTO abstract science, as we have before observed, the notion of cause does not enter. The truths it is conversant with are *necessary* ones, and exist independent of cause. There may be no such real *thing* as a right-lined triangle marked out in space; but the moment we conceive one in our minds, we cannot refuse to admit the sum of its three angles to be equal to two right angles; and if, in addition, we conceive one of its angles to be a right angle, we cannot thenceforth refuse to admit that the sum of the squares on the two sides, including the right angle, is equal to the square on the side subtending it. To maintain the contrary, would be, in effect, to deny its being right-angled. No one *causes* or *makes* all the diameters of an ellipse to be bisected in its centre. To assert the contrary, would not be to rebel against a power, but to deny our own words. But in natural science, *cause* and *effect* are the ultimate relations we contemplate; and *laws*, whether imposed or maintained, which, for aught we can perceive, might have been other than they are. This distinction is very important. A clever man, shut up alone and allowed unlimited time, might reason out for himself all the truths of mathematics, by proceeding from those simple notions of space and number of which he cannot divest himself without ceasing to think. But he could never tell, by any effort of reasoning, what would become of a lump of sugar if immersed in water, or what impression would be produced on his eye by mixing the colors yellow and blue.

We have thus pointed out to us, as the great, and indeed only ultimate source of our knowledge of nature and its laws, EXPERIENCE; by which we mean, not the experience of one man only, or of one generation, but the accumulated experience of all mankind, in all ages, registered in books or recorded by tradition. But experience may be acquired in two ways: either, first, by noticing facts as they occur, without any

DOI: 10.4324/9781003009337-6

attempt to influence the frequency of their occurrence, or to vary the circumstances under which they occur; this is OBSERVATION: or, secondly, by putting in action causes and agents over which we have control, and purposely varying their combinations, and noticing what effects take place; this is EXPERIMENT. To these two sources we must look as the fountains of all natural science. It is not intended, however, by thus distinguishing observation from experiment, to place them in any kind of contrast. Essentially they are much alike, and differ rather in degree than in kind; so that, perhaps, the terms *passive* and *active observation* might better express their distinction; but it is, nevertheless, highly important to mark the different states of mind in inquiries carried on by their respective aids, as well as their different effects in promoting the progress of science. In the former, we sit still and listen to a tale, told us, perhaps obscurely, piecemeal, and at long intervals of time, with our attention more or less awake. It is only by after-rumination that we gather its full import; and often, when the opportunity is gone by, we have to regret that our attention was not more particularly directed to some point which, at the time, appeared of little moment, but of which we at length appreciate the importance. In the latter, on the other hand, we cross-examine our witness, and, by comparing one part of his evidence with the other, while he is yet before us, and reasoning upon it in his presence, are enabled to put pointed and searching questions, the answer to which may at once enable us to make up our minds. Accordingly it has been found invariably, that in those departments of physics where the phenomena are beyond our control, or into which experimental inquiry, from other causes, has not been carried, the progress of knowledge has been slow, uncertain, and irregular; while in such as admit of experiment, and in which mankind have agreed to its adoption, it has been rapid, sure, and steady. For example, in our knowledge of the nature and causes of volcanoes, earthquakes, the fall of stones from the sky, the appearance of new stars and disappearance of old ones, and other of those great phenomena of nature which are altogether beyond our command, and at the same time are of too rare occurrence to permit any one to repeat and rectify his impressions respecting them, we know little more now than in the earliest times. Here our tale is told us slowly, and in broken sentences. In astronomy, again, we have at least an uninterrupted narrative; the opportunity of observation is constantly present, and makes up in some measure for the impossibility of varying our point of view, and calling for information at the precise moment it is wanted. Accordingly, astronomy, regarded as a science of mere observation, arrived, though by very slow degrees, to a state of considerable maturity. But the moment that it became a branch of mechanics, a science essentially experimental (that is to say, one in which any principle laid down can be subjected to immediate and decisive *trial*, and where experience does not require to be waited for), its progress suddenly acquired a tenfold acceleration; nay, to such a degree, that it has been asserted, and we believe with truth, that were the records of all observations from the earliest ages annihilated, leaving only those made in a single observatory,[1] during a single lifetime,[2] the whole of this most perfect of sciences might, from those data, and as to the objects included in them, be at once reconstructed, and appear precisely as it stood at their conclusion. To take another instance: mineralogy, till modern times, could hardly be said to exist. The description of even the precious stones in Theophrastus and Pliny are, in most cases, hardly sufficient to identify them, and in many fall short even of that humble object: more recent observers, by attending more carefully to the obvious characters of minerals, had formed a pretty extensive catalogue of them, and made various attempts to arrange and methodize the knowledge thus acquired, and even to deduce some general conclusions respecting the forms they habitually assume: but from the moment that chemical analysis was applied to resolve them into their constituent elements, and that, led by a happy accident, the genius of Haüy discovered the general fact, that they could be *cloven* or split in such directions as to lay bare their peculiar primitive or fundamental forms (which lay concealed within them, as the statue might be conceived encrusted in its marble envelope),—from that moment, mineralogy ceased to be an unmeaning list of names, a mere laborious cataloguing of stones and rubbish, and became, what it now is, a regular, methodical, and most important science, in which every year is bringing to light new relations, new laws, and new practical applications.

Experience once recognised as the fountain of all our knowledge of nature, it follows that, in the study of nature and its laws, we ought at once to make up our minds to dismiss as idle prejudices, or at least suspend as premature, any preconceived notion of what might or what ought to be the order of nature in any proposed case, and content ourselves with observing, as a plain matter of fact, what *is*. To experience we refer, as the only ground of all physical inquiry. But before experience itself can be used with advantage, there is one preliminary step to make, which depends wholly on ourselves: it is the absolute dismissal and clearing the mind of all prejudice, from whatever source arising, and the determination to stand and fall by the result of a direct appeal to facts in the first instance, and of strict logical deduction from them afterwards. Now, it is necessary to distinguish between two kinds of prejudices, which exercise very different dominion over the mind, and, moreover, differ extremely in the difficulty of dispossessing them, and the process to be gone through for that purpose. These are,—

1. Prejudices of opinion.
2. Prejudices of sense.

By prejudices of opinion, we mean opinions hastily taken up, either from the assertion of others, from our own superficial views, or from vulgar observation, and which, from being constantly admitted without dispute, have obtained the strong hold of habit on our minds. Such were the opinions once maintained that the earth is the greatest body in the universe, and placed immovable in its centre, and all the rest of the universe created for its sole use; that it is the nature of fire and of sounds to ascend; that the moonlight is cold; that dews *fall* from the air, &c.

To combat and destroy such prejudices we may proceed in two ways, either by demonstrating the falsehood of the facts alleged in their support, or by showing how the appearances, which seem to countenance them, are more satisfactorily accounted for without their admission. But it is unfortunately the nature of prejudices of opinion to adhere, in a certain degree, to every mind, and to some with pertinacious obstinacy, *pigris radicibus*, after all ground for their reasonable entertainment is destroyed. Against such a disposition the student of natural science must contend with all his power. Not that we are so unreasonable as to demand of him an instant and peremptory dismission of all his former opinions and judgments; all we require is, that he will hold them without bigotry, retain till he shall see reason to question them, and be ready to resign them when fairly proved untenable, and to doubt them when the weight of probability is shown to lie against them. If he refuse this, he is incapable of science.

Our resistance against the destruction of the other class of prejudices, those of sense, is commonly more violent at first, but less persistent, than in the case of those of opinion. Not to trust the evidence of our senses seems indeed a hard condition, and one which, if proposed, none would comply with. But it is not the direct evidence of our senses that we are in any case called upon to reject, but only the erroneous judgments we unconsciously form from them, and this only when they can be shown to be so *by counter evidence of the same sort;* when one sense is brought to testify against another, for instance; or the same sense against itself, and the obvious conclusions in the two cases disagree, so as to compel us to acknowledge that one or other must be wrong. For example, nothing at first can seem a more rational, obvious, and incontrovertible conclusion, than that the *color* of an object is an inherent quality, like its weight, hardness, &c., and that to *see* the object, and see it *of its own color*, when nothing intervenes between our eyes and it, are one and the same thing. Yet this is only a prejudice; and that it is so, is shown by bringing forward the same sense of vision which led to its adoption, as evidence on the other side; for, when the differently colored prismatic rays are thrown, in a dark room, in succession upon any object, whatever be the color we are in the habit of calling its own, it will appear of the particular hue of the light which falls upon it: a yellow paper, for instance, will appear scarlet when illuminated by red rays, yellow when by yellow, green by green, and blue by blue rays; its own (so called) proper color *not in the least degree mixing with that it so exhibits*.

To give one or two more examples of the kind of illusion which the senses practise on us, or rather which we practise on ourselves, by a misinterpretation of their evidence: the moon at its rising and setting appears much larger than when high up in the sky. This is, however, a mere erroneous judgment; for when we come to measure its diameter, so far from finding our conclusion borne out by fact, we actually find it to measure materially less. Here is eyesight opposed to eyesight, with the advantage of deliberate measurement. In ventriloquism we have the hearing at variance with all the other senses, and especially with the sight, which is sometimes contradicted by it in a very extraordinary and surprising manner, as when the voice is made to seem to issue from an inanimate and motionless object. If we plunge our hands, one into ice-cold water, and the other into water as hot as can be borne, and, after letting them stay awhile, suddenly transfer them both to a vessel full of water at a blood heat, the one will feel a sensation of heat, the other of cold. And if we cross the two first fingers of one hand, and place a pea in the fork between them, moving and rolling it about on a table, we shall (especially if we close our eyes) be fully persuaded we have two peas. If the nose be held while we are eating cinnamon, we shall perceive no difference between its flavor and that of a deal shaving.

These, and innumerable instances we might cite, will convince us, that though we are never deceived in the *sensible impression* made by external objects on us, yet in forming our judgments of them we are greatly at the mercy of circumstances, which either modify the impressions actually received, or combine them, with adjuncts which have become habitually associated with different judgments; and, therefore, that, in estimating the degree of confidence we are to place in our conclusions, we must, of necessity, take into account these modifying or accompanying circumstances, whatever they may be. We do not, of course, here speak of deranged organization; such as, for instance, a distortion of the eye, producing double vision, and still less of mental delusion, which absolutely perverts the meaning of sensible impressions.

As the mind exists not in the place of sensible objects, and is not brought into immediate relation with them, we can only regard sensible impressions as signals conveyed from them by a wonderful, and, to us, inexplicable mechanism, to our minds, which receives and reviews them, and, by habit and association, connects them with corresponding qualities or affections in the objects; just as a person writing down and comparing the signals of a telegraph might interpret their meaning. As, for instance, if he had constantly observed that the exhibition of a certain signal was sure to be followed next day by the announcement of the arrival of a ship at Portsmouth, he would connect the two facts by a link of the very same nature with that which connects the notion of a large wooden building, filled with sailors, with the impression of her outline on the retina of a spectator on the beach.

In captain Head's amusing and vivid description of his journey across the Pampas of South America occurs an anecdote quite in point. His guide one day suddenly stopped him, and, pointing high into the air, cried out, "A lion!" Surprised at such an exclamation, accompanied with such an act, he turned up his eyes, and with difficulty perceived, at an immeasurable height, a flight of condors soaring in circles in a particular spot. Beneath that spot, far out of sight of himself or guide, lay the carcass of a horse, and over that carcass stood (as the guide well knew) the lion, whom the condors were eyeing with envy from their airy height. The signal of the birds was to him what the sight of the lion alone could have been to the traveller—a full assurance of its existence.

Notes

1 Greenwich.
2 Maskelyne's.

Part 2

MEDICINE

Editorial Headnote

Edward Jenner, *An Inquiry into the Causes and Effects of the Variolae Vaccinae: A Disease Discovered in Some of the Western Counties of England, Particularly Gloucestershire, and Known by the Name of the Cow Pox* (Private printing, London: first published, 1798, this edition, 1800), 9–36

Edward Jenner, 1749–1823, was a Gloucestershire physician. Others have laid claim to the discovery of cowpox and its relation to smallpox, but Jenner was the first systematically to test arm-to-arm inoculation of cowpox among humans and to check its protective effect against smallpox. The experiments form part of a life of curious tinkering in medical and scientific problems, which included a years-long failure to understand the nature of tuberculosis and a certain fame deriving from his discovery of the life-cycle of the cuckoo.

4

AN INQUIRY INTO THE CAUSES AND EFFECTS OF THE VARIOLAE VACCINAE

A Disease Discovered in Some of the Western Counties of England, Particularly Gloucestershire, and Known by the Name of the Cow Pox

Edward Jenner

Source: Private printing, London: first published, 1798, this edition, 1800, 9–36

CASE I

JOSEPH MERRET, now an Under Gardener to the Earl of Berkeley, lived as a Servant with a Farmer near this place in the year 1770, and occasionally assisted in milking his master's cows. Several horses belonging to the farm began to have sore heels, which Merret frequently attended. The cows soon became affected with the Cow Pox, and soon after several sores appeared on his hands. Swellings and stiffness in each axilla followed, and he was so much indisposed for several days as to be incapable of pursuing his ordinary employment. Previously to the appearance of the distemper among the cows there was no fresh cow brought into the farm, nor any servant employed who was affected with the Cow Pox.

In April, 1795, a general inoculation taking place here, Merret was inoculated with his family; so that a period of twenty-five years had elapsed from his having the Cow Pox to this time. However, though the variolous matter was repeatedly inserted into his arm, I found it impracticable to infect him with it; an efflorescence only, taking on an erysipelatous look about the centre, appearing on the skin near the punctured parts. During the whole time that his family had the Small Pox, one of whom had it very full, he remained in the house with them, but received no injury from exposure to the contagion.

It is necessary to observe, that the utmost care was taken to ascertain, with the most scrupulous precision, that no one whose case is here adduced had gone through the Small Pox previous to these attempts to produce that disease.

DOI: 10.4324/9781003009337-8

Had these experiments been conducted in a large city, or in a populous neighbourhood, some doubts might have been entertained; but here, where population is thin, and where such an event as a person's having had the Small Pox is always faithfully recorded, no risk of inaccuracy in this particular can arise.

CASE II

SARAH PORTLOCK, of this place, was infected with the Cow Pox, when a Servant at a Farmer's in the neighbourhood, twenty-seven years ago.[1]

In the year 1792, conceiving herself, from this circumstance, secure from the infection of the Small Pox, she nursed one of her own children who had accidentally caught the disease, but no indisposition ensued.—During the time she remained in the infected room, variolous matter was inserted into both her arms, but without any further effect than in the preceding case.

CASE III

JOHN PHILLIPS, a Tradesman of this town, had the Cow Pox at so early a period as nine years of age. At the age of sixty-two I inoculated him, and was very careful in selecting matter in its most active state. It was taken from the arm of a boy just before the commencement of the eruptive fever, and instantly inserted. It very speedily produced a sting-like feel in the part. An efflorescence appeared, which on the fourth day was rather extensive, and some degree of pain and stiffness were felt about the shoulder; but on the fifth day these symptoms began to disappear, and in a day or two after went entirely off, without producing any effect on the system.

CASE IV

MARY BARGE, of Woodford, in this parish, was inoculated with variolous matter in the year 1791. An efflorescence of a palish red colour soon appeared about the parts where the matter was inserted, and spread itself rather extensively, but died away in a few days without producing any variolous symptoms.[2] She has since been repeatedly employed as a nurse to Small-pox patients, without experiencing any ill consequences. This woman had the Cow Pox when she lived in the service of a Farmer in this parish thirty-one years before.

CASE V

MRS. H———, a respectable Gentlewoman of this town, had the Cow Pox when very young. She received the infection in rather an uncommon manner: it was given by means of her handling some of the same utensils[3] which were in use among the servants of the family, who had the disease from milking infected cows. Her hands had many of the Cow-pox sores upon them, and they were communicated to her nose, which became inflamed and very much swoln. Soon after this event Mrs. H—— was exposed to the contagion of the Small Pox, where it was scarcely possible for her to have escaped, had she been susceptible of it, as she regularly attended a relative who had the disease in so violent a degree that it proved fatal to him.

In the year 1778 the Small Pox prevailed very much at Berkeley, and Mrs. H—— not feeling perfectly satisfied respecting her safety (no indisposition having followed her exposure to the Small Pox) I inoculated her with active variolous matter. The same appearance followed as in the preceding cases—an efflorescence on the arm without any effect on the constitution.

CASE VI

IT is a fact so well known among our Dairy Farmers, that those who have had the Small Pox either escape the Cow Pox or are disposed to have it slightly; that as soon as the complaint shews itself among the cattle, assistants are procured, if possible, who are thus rendered less susceptible of it, otherwise the business of the farm could scarcely go forward.

In the month of May, 1796, the Cow Pox broke out at Mr. Baker's, a Farmer who lives near this place. The disease was communicated by means of a cow which was purchased in an infected state at a neighbouring fair, and not one of the Farmer's cows (consisting of thirty) which were at that time milked escaped the contagion. The family consisted of a man servant, two dairymaids, and a servant boy, who, with the Farmer himself, were twice a day employed in milking the cattle. The whole of this family, except Sarah Wynne, one of the dairymaids, had gone through the Small Pox. The consequence was, that the Farmer and the servant boy escaped the infection of the Cow Pox entirely, and the servant man and one of the maid servants had each of them nothing more than a sore on one of their fingers, which produced no disorder in the system. But the other dairymaid, Sarah Wynne, who never had the Small Pox, did not escape in so easy a manner. She caught the complaint from the cows, and was affected with the symptoms described in the 5th page in so violent a degree, that she was confined to her bed, and rendered incapable for several days of pursuing her ordinary vocations in the farm.

March 28th, 1797, I inoculated this girl, and carefully rubbed the variolous matter into two slight incisions made upon the left arm. A little inflammation appeared in the usual manner around the parts where the matter was inserted, but so early as the fifth day it vanished entirely without producing any effect on the system.

CASE VII

ALTHOUGH the preceding history pretty clearly evinces that the constitution is far less susceptible of the contagion of the Cow Pox after it has felt that of the Small Pox, and although in general, as I have observed, they who have had the Small Pox, and are employed in milking cows which are infected with the Cow Pox, either escape the disorder, or have sores on the hands without feeling any general indisposition, yet the animal economy is subject to some variation in this respect, which the following relation will point out:

In the summer of the year 1796 the Cow Pox appeared at the Farm of Mr. Andrews, a considerable dairy adjoining to the town of Berkeley. It was communicated, as in the preceding instance, by an infected cow purchased at a fair in the neighbourhood. The family consisted of the Farmer, his wife, two sons, a man and a maid servant; all of whom, except the Farmer (who was fearful of the consequences), bore a part in milking the cows. The whole of them, exclusive of the man servant, had regularly gone through the Small Pox; but in this case no one who milked the cows escaped the contagion. All of them had sores upon their hands, and some degree of general indisposition, preceded by pains and tumours in the axillæ: but there was no comparison in the severity of the disease as it was felt by the servant man, who had escaped the Small Pox, and by those of the family who had not, for, while he was consined to his bed, they were able, without much inconvenience, to follow their ordinary business.

February the 13th, 1797, I availed myself of an opportunity of inoculating William Rodway, the servant man above alluded to. Variolous matter was inserted into both his arms; in the right by means of superficial incisions, and into the left by slight punctures into the cutis. Both were perceptibly inflamed on the third day. After this the inflammation about the punctures soon died away, but a small appearance of erysipelas was manifest about the edges of the incisions till the eighth day, when a little uneasiness was felt for the space of half an hour in the right axilla. The inflammation then hastily disappeared without producing the most distant mark of affection of the system.

CASE VIII

ELIZABETH WYNNE, aged fifty-seven, lived as a servant with a neighbouring Farmer thirty-eight years ago. She was then a dairymaid, and the Cow Pox broke out among the cows. She caught the disease with the rest of the family, but, compared with them, had it in a very slight degree, one very small sore only breaking out on the little finger of her left hand, and scarcely any perceptible indisposition following it.

As the malady had shewn itself in so slight a manner, and as it had taken place at so distant a period of her life, I was happy with the opportunity of trying the effects of variolous matter upon her constitution, and on the 28th of March, 1797, I inoculated her by making two superficial incisions on the left arm, on which the matter was cautiously rubbed. A little efflorescence soon appeared, and a tingling sensation was felt about the parts where the matter was inserted until the third day, when both began to subside, and so early as the fifth day it was evident that no indisposition would follow.

CASE IX

ALTHOUGH the Cow Pox shields the constitution from the Small Pox, and the Small Pox proves a protection against its own future poison, yet it appears that the human body is again and again susceptible of the infectious matter of the Cow Pox, as the following history will demonstrate:

William Smith, of Pyrton in this parish, contracted this disease when he lived with a neighbouring Farmer in the year 1780. One of the horses belonging to the farm had sore heels, and it fell to his lot to attend him. By these means the infection was carried to the cows, and from the cows it was communicated to Smith. On one of his hands were several ulcerated sores, and he was affected with such symptoms as have been before described.

In the year 1791 the Cow Pox broke out at another farm where he then lived as a servant, and he became affected with it a second time; and in the year 1794 he was so unfortunate as to catch it again. The disease was equally as severe the second and third time as it was on the first.[4]

In the spring of the year 1795 he was twice inoculated, but no affection of the system could be produced from the variolous matter; and he has since associated with those who had the Small Pox in its most contagious state without feeling any effect from it.

CASE X

SIMON NICHOLS lived as a servant with Mr. Bromedge, a gentleman who resides on his own farm in this parish, in the year 1782. He was employed in applying dressings to the sore heels of one of his master's horses, and at the same time assisted in milking the cows. The cows became affected in consequence, but the disease did not shew itself on their nipples till several weeks after he had begun to dress the horse. He quitted Mr. Bromedge's service, and went to another farm without any sores upon him; but here his hands soon began to be affected in the common way, and he was much indisposed with the usual symptoms. Concealing the nature of the malady from Mr. Cole, his new master, and being there also employed in milking, the Cow Pox was communicated to the cows.

Some years afterwards Nichols was employed in a farm where the Small Pox broke out, when I inoculated him with several other patients, with whom he continued during the whole time of their confinement. His arm inflamed, but neither the inflammation nor his associating with the inoculated family produced the least effect upon his constitution.

CASE XI

WILLIAM STINCHCOMB was a fellow servant with Nichols at Mr. Bromedge's Farm at the time the cattle had the Cow Pox, and he was unfortunately infected by them. His left hand was very severely affected with several corroding ulcers, and a tumour of considerable size appeared in the axilla of that side. His right hand had only one small sore upon it, and no tumour discovered itself in the corresponding axilla.

In the year 1792 Stinchcomb was inoculated with variolous matter, but no consequences ensued beyond a little inflammation in the arm for a few days. A large party were inoculated at the same time, some of whom had the disease in a more violent degree than is commonly seen from inoculation. He purposely associated with them, but could not receive the Small Pox.

During the sickening of some of his companions, their symptoms so strongly recalled to his mind his own state when sickening with the Cow Pox, that he very pertinently remarked their striking similarity.

CASE XII

THE Paupers of the village of Tortworth, in this county, were inoculated by Mr. Henry Jenner, Surgeon, of Berkeley, in the year 1795. Among them, eight patients presented themselves who had at different periods of their lives had the Cow Pox. One of them, Hester Walkley, I attended with that disease when she lived in the service of a Farmer in the same village in the year 1782; but neither this woman, nor any other of the patients who had gone through the Cow Pox, received the variolous infection either from the arm or from mixing in the society of the other patients who were inoculated at the same time. This state of security proved a fortunate circumstance, as many of the poor women were at the same time in a state of pregnancy.

CASE XIII

One instance has occurred to me of the system being affected from the matter issuing from the heels of horses, and of its remaining afterwards unsusceptible of the variolous contagion; another, where the Small Pox appeared obscurely; and a third, in which its complete existence was positively ascertained.

First, THOMAS PEARCE, is the son of a Smith and Farrier near to this place. He never had the Cow Pox; but, in consequence of dressing horses with sore heels at his father's, when a lad, he had sores on his fingers which suppurated, and which occasioned a pretty severe indisposition. Six years afterwards I inserted variolous matter into his arm repeatedly, without being able to produce any thing more than slight inflammation, which appeared very soon after the matter was applied, and afterwards I exposed him to the contagion of the Small Pox with as little effect.[5]

CASE XIV

Secondly, Mr. JAMES COLE, a Farmer in this parish, had a disease from the same source as related in the preceding case, and some years after was inoculated with variolous matter. He had a little pain in the axilla, and felt a slight indisposition for three or four hours. A few eruptions shewed themselves on the forehead, but they very soon disappeared without advancing to maturation.

CASE XV

Although in the two former instances the system seemed to be secured, or nearly so, from variolous infection, by the absorption of matter from sores produced by the diseased heels of horses, yet the following case decisively proves that this cannot be entirely relied upon, until a disease has been generated by the morbid matter from the horse on the nipple of the cow, and passed through that medium to the human subject.

Mr. ABRAHAM RIDDIFORD, a Farmer at Stone in this parish, in consequence of dressing a mare that had sore heels, was affected with very painful sores in both his hands, tumours in each axilla, and severe and general indisposition. A Surgeon in the neighbourhood attended him, who, knowing the similarity between the appearance of the sores upon his hands and those produced by the Cow Pox, and being acquainted also with the effects of that disease on the human constitution, assured him that he never need to fear the infection of the Small Pox; but this assertion proved fallacious, for, on being exposed to the infection upwards of twenty years afterwards, he caught the disease, which took its regular course in a very mild way. There certainly was a difference perceptible, although it is not easy to describe it, in the general appearance of the pustules from that which we commonly see. Other practitioners, who visited the patient at my request, agreed with me in this point, though there was no room left for suspicion as to the reality of the disease, as I inoculated some of his family from the pustules, who had the Small Pox, with its usual appearances, in consequence.

CASE XVI

SARAH NELMES, a dairymaid at a Farmer's near this place, was infected with the Cow Pox from her master's cows in May, 1796. She received the infection on a part of the hand which had been previously in a slight degree injured by a scratch from a thorn. A large pustulous sore and the usual symptoms accompanying the disease were produced in consequence. The pustule was so expressive of the true character of the Cow Pox, as it commonly appears upon the hand, that I have given a representation of it in the annexed plate. The two small pustules on the wrists arose also from the application of the virus to some minute abrasions of the cuticle, but the livid tint, if they ever had any, was not conspicuous at the time I saw the patient. The pustule on the fore finger shews the disease in an earlier stage. It did not actually appear on the hand of this young woman, but was taken from that of another, and is annexed for the purpose of representing the malady after it has newly appeared.

CASE XVII

THE more accurately to observe the progress of the infection, I selected a healthy boy, about eight years old, for the purpose of inoculation for the Cow Pox. The matter was taken from a sore on the hand of a dairymaid,[6] who was infected by her master's cows, and it was inserted, on the 14th of May, 1796, into the arm of the boy by means of two superficial incisions, barely penetrating the cutis, each about half an inch long.

On the seventh day he complained of uneasiness in the axilla, and on the ninth he became a little chilly, lost his appetite, and had a slight head-ach. During the whole of this day he was perceptibly indisposed, and spent the night with some degree of restlessness, but on the day following he was perfectly well.

The appearance of the incisions in their progress to a state of maturation were much the same as when produced in a similar manner by variolous matter. The only difference which I perceived was, in the state of the limpid fluid arising from the action of the virus, which assumed rather a darker hue, and in that of the efflorescence spreading round the incisions, which had more of an erysipelatous look than we commonly perceive when variolous matter has been made use of in the same manner; but the whole died away (leaving on the inoculated parts scabs and subsequent eschars) without giving me or my patient the least trouble.

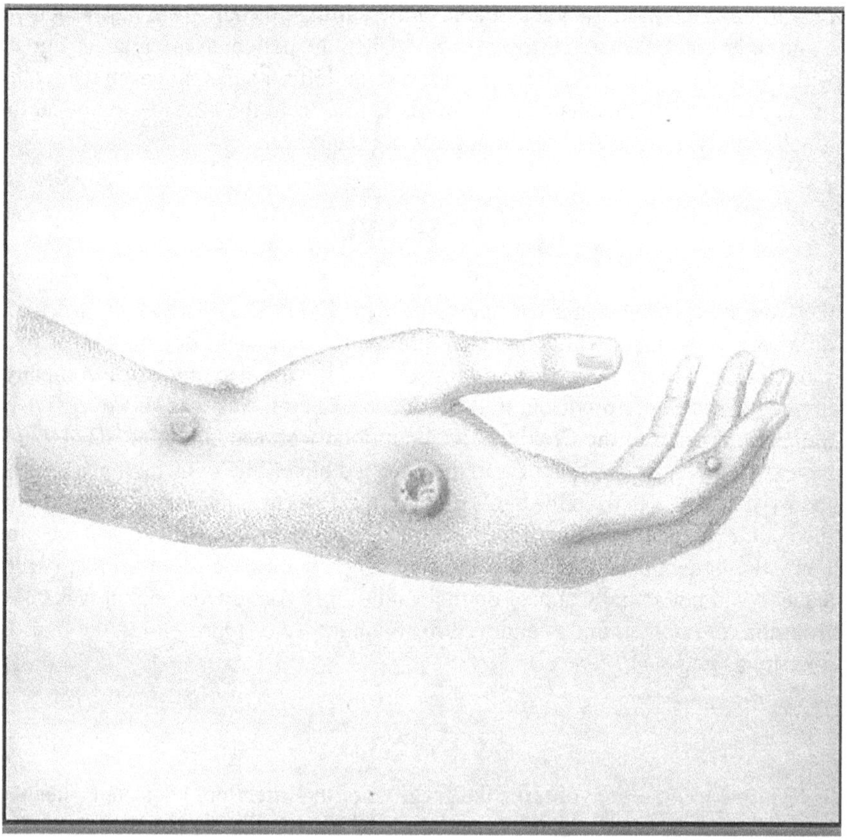

In order to ascertain whether the boy, after feeling so slight an affection of the system from the Cow-pox virus, was secure from the contagion of the Small-pox, he was inoculated the ist of July following with variolous matter, immediately taken from a pustule. Several slight punctures and incisions were made on both his arms, and the matter was carefully inserted, but no disease followed. The same appearances were observable on the arms as we commonly see when a patient has had variolous matter applied, after having either the Cow-pox or the Small-pox. Several months afterwards, he was again inoculated with variolous matter, but no sensible effect was produced on the constitution.

Here my researches were interrupted till the spring of the year 1798, when from the wetness of the early part of the season, many of the farmers' horses in this neighbourhood were affected with sore heels, in consequence of which the Cow-pox broke out among several of our dairies, which afforded me an opportunity of making further observations upon this curious disease.

A mare, the property of a person who keeps a dairy in a neighbouring parish, began to have sore heels the latter end of the month of February 1798, which were occasionally washed by the servant men of the farm, Thomas Virgoe, William Wherret, and William Haynes, who in consequence became affected with sores in their hands, followed by inflamed lymphatic glands in the arms and axillæ, shiverings succeeded by heat, lassitude and general pains in the limbs. A single paroxysm terminated the disease; for within twenty-four hours they were free from general indisposition, nothing remaining but the sores on their hands. Haynes and Virgoe, who had gone through the Small-pox from inoculation, described their feelings as very similar to those which affected them on sickening with that malady. Wherret never had had the Small-pox. Haynes was daily employed as one of the milkers at the farm, and the disease began to shew itself among the cows about ten days after he first assisted in washing the mare's heels. Their nipples became sore in the usual way, with blueish pustules; but as remedies were early applied they did not ulcerate to any extent.

Notes

1 I have purposely selected several cases in which the disease had appeared at a very distant period previous to the experiments made with variolous matter, to shew that the change produced in the constitution is not affected by time.

2 It is remarkable that variolous matter, when the system is disposed to reject it, should excite inflammation on the part to which it is applied more speedily than when it produces the Small Pox. Indeed it becomes almost a criterion by which we can determine whether the infection will be received or not. It seems as if a change, which endures through life, had been produced in the action, or disposition to action, in the vessels of the skin; and it is remarkable too, that whether this change has been effected by the Small Pox, or the Cow Pox, that the disposition to sudden cuticular inflammation is the same on the application of variolous matter.

3 When the Cow Pox has prevailed in the dairy, it has often been communicated to those who have not milked the cows, by the handle of the milk pail.

4 This is not the case in general—a second attack is commonly very slight, and so, I am informed, it is among the cows.

5 It is a remarkable fact, and well known to many, that we are frequently foiled in our endeavours to communicate the Small Pox by inoculation to blacksmiths, who in the country are farriers. They often, as in the above instance, either resist the contagion entirely, or have the disease anomalously. Shall we not be able now to account for this on a rational principle?

6 From the sore on the hand of Sarah Nelmes.—See the preceding case and the plate.

Editorial Headnote

Emma Hardinge Britten, *The Electric Physician: Or Self Cure Through Electricity* (William Britten: Boston, 1875), 3–6

Emma Hardinge Britten, 1823–1899, was a forerunner in the spiritualist movement, first in England, later in the United States. She was not considered a scientist, but this work made direct appeals to science and medicine and proclaimed a practical experience with the techniques of electrical cure.

5

THE ELECTRIC PHYSICIAN
Or Self Cure Through Electricity

Emma Hardinge Britten

Source: William Britten: Boston, 1875, 3–6

THE object of this little work is to supply to the heads of families and non-professional persons, the means of applying for themselves the best method of cure yet discovered, and to suggest to professional students and medical practitioners, how they can acquire the elements of a beneficent and invaluable remedial art, which at present forms no part of a collegiate education.

Many learned treatises on electricity are already before the world, but few if any purely practical methods have as yet been published which would enable the unscientific reader to apply electricity to the cure of disease on safe, reliable and scientific principles.

The essays put forth under the attractive guise of information on the uses of electricity as a remedial agent consist for the most part of details concerning the history of discoveries and discoverers, experiments and experimenters, theories which the experiences of every fresh decade of time will be sure to overthrow, and abstractions which can serve no other possible use than to fill up so many pages of a *salable* volume.

When the author has been questioned by her friends, patients or pupils, what work they could consult that would enable them to utilize electricity as a reliable remedial agent in family practice, she has felt keenly the necessity for some treatise or guide devoted solely to this purpose, and free from the mass of irrelevant matter with which the literature of this subject is overladen, and it is with a view to meet this much needed desideratum that the following unpretentious pages have been written.

The author does not deem it necessary to inform the reader how or when the uses of electricity were first discovered. No learned disquisition on the origin of its nomenclature is required to instruct a person suffering from a raging toothache how they can utilize electricity for its relief, nor need the hapless victim of a severe neuralgic attack, study out the names of all the great or little personages who have experimented with electricity, before they can apply it for their own benefit.

DOI: 10.4324/9781003009337-9

Neither does it seem fitting to preface these pages, or subtract one iota from their singleness of purpose, by introducing the elements of anatomy and physiology, a course so universally adopted by medical writers however special their subject may be, that we can now scarcely pick up from our area steps a printed handbill inviting sufferers of all classes to go and be cured by some universal nostrum without being treated to a full and comprehensive detail of the human system in all its tissues, organs and parts.

Whilst the labors which a Wilson or a Carpenter have spread over six hundred pages of a medical treatise, can thus be crowded up into one sheet of *popular literature*, whilst every book-stall abounds with anatomical knowledge "for the million," and every common school affords to the rising generation ample means of studying the generalities of physiological law, it would be a reproach to the intelligence of the age to suppose that any readers of this work can be wholly ignorant of such important subjects. Assuming then that those whom we address are at least fairly educated persons, that being so, they must be familiar with the general principles of anatomical structure and physiological action, and that their chief aim in the perusal of these pages is to learn what electricity can do for them in the way of alleviating suffering, and supplementing the imperfect text-books of so-called modern "medical science," we shall at once proceed to give plain and comprehensible directions for the application of electricity as the latest discovered and best method of remedial art now known to man.

Editorial Headnote

Walter Bradford Cannon, *A Laboratory Course in Physiology* (1911; 4th edn, Cambridge: Harvard University Press, 1923), xv–xvi

Walter Bradford Cannon, 1871–1945, was a giant figure in American medical science, leading Harvard Medical School's Department of Physiology, and spearheading the defence of experimental medicine against antivivisection campaigns. Here he lays out the importance of direct experience.

6

A LABORATORY COURSE IN PHYSIOLOGY

Walter Bradford Cannon

Source: 1911; 4th edn, Cambridge: Harvard University Press, 1923, xv–xvi

To the student of biology and the student of that part of applied biology included in medicine, the laboratory method of learning the functions of the tissues and organs of the animal body has demonstrated its superiority over other methods, not only in impressing vividly the facts of observation, but also in demanding a discipline which furnishes to the student power to go forth with assurance in independent search.

That the student may learn the facts of the science important for his further progress, experiments must be selected which have for that purpose the most significance. That the student may receive the greatest benefit from the discipline of his study, it is essential that he be placed so far as possible on his own resources. He should be required, under supervision, to make and record his own observations, and draw his own conclusions, and should be urged at every step to suggest further observation or experiment to test those conclusions. A student who has conscientiously applied his abilities under such requirements possesses at the end a first-hand acquaintance with the subject-matter of physiology, and has formed besides a very valuable set of habits – the habits of careful observation, of concise and accurate statement of things observed, and of exact inference from these observations. He may also have been stimulated to exercise his imagination in the planning of further experimentation to discover new knowledge.

That the purposes of laboratory instruction may be served, the attempt has been made to arrange the experiments in this book, in both the larger and smaller divisions of the subject, so as to render answers in an orderly manner to the following questions: –

First. Of what activity is an organ or tissue capable?

Second. How do certain factors condition or modify that activity?

An attempt has been made also to arrange the experiments so that facts with widest bearing are observed first. Such a plan permits a statement of subsequent experiments in a manner which will call forth the greatest use of the student's intelligence – he is required repeatedly to rely on his previous observations.

DOI: 10.4324/9781003009337-10

Much importance of the facts of observation lies in their relations to one another, that is, in the generalizations which may be drawn after comparing them. To stimulate students to regard thoughtfully what they see and to note the relation of present phenomena to those already observed, questions and suggestions have been found valuable. They tend to prevent the laboratory exercise from degenerating into an automatic following of directions in the book, and an automatic recording of what the book states will occur. The student should consider carefully any questions given in the text or suggested during the course of experimentation.

It is important for the student to realize that in performing his experiments he is following the ancient way which led to knowledge – in short, that he is learning the methods of progress by repeating the procedures of the original discoverers. In order to make this relation clear, references are frequently given to the first publication of the method and the results. In some instances, also, references are made to related literature, which will permit the student to become better acquainted with the significance of his work, in case his interest is aroused.

The selection of experiments has been made on the assumption that students will have additional practical experience in biological chemistry.

To save the student's time and to afford him at the end of the course a concise and convenient account of his laboratory experiences, the descriptions of experiments have been printed only on the left-hand pages in order that the notes may be written opposite on the right-hand pages. When a graphic record is made it should be preserved with the notes on observations. The paper bearing the record should be carefully trimmed, then pasted in the book and protected by tissue-paper from disfigurement by rubbing. From the beginning the student should plan to record his results in an orderly manner, giving the conditions of the experiment, the observations made, and the inference, if an inference may properly be drawn. The notes should be written as soon as the experiment is finished. They should be brief and clear. Diagrams often serve better than verbal descriptions.

The directions for experimentation here presented are the result of numerous corrections and alterations, based on experience with large classes. I wish to express my gratitude to Dr. E. G. Martin, Dr. E. B. Meigs, and Dr. R. G. Hoskins for many valuable suggestions in the earlier development of the "Course," and especially to Dr. C. K. Drinker and Dr. A. C. Redfield for important contributions to later editions.

Part 3

GALVANISM

Editorial Headnote

John Aldini, *An Account of the Galvanic Experiments Performed by John Aldini on the Body of a Malefactor* (London: Cuthell and Martin, 1803), 1–34

Giovanni Aldini, 1762–1834, was a professor of experimental physics at the University of Bologna. His work on the science of galvanism followed the innovations of his predecessor and relative, Luigi Galvani, and caused a stir throughout Europe as the limbs of the dead were re-animated.

7

AN ACCOUNT OF THE GALVANIC EXPERIMENTS PERFORMED BY JOHN ALDINI ON THE BODY OF A MALEFACTOR

John Aldini

Source: London: Cuthell and Martin, 1803, 1–34

THE unenlightened part of mankind are apt to entertain a prejudice against those, however laudable their motives, who attempt to perform experiments on dead subjects; and the vulgar in general even attach a sort of odium to the common practice of anatomical dissection. It is, however, an incontrovertible fact, that such researches in modern times have proved a source of the most valuable information, in regard to points highly interesting to the knowledge of the human frame, and have contributed in an eminent degree to the improvement of physiology and anatomy. Enlightened legislators have been sensible of this truth; and therefore it has been wisely ordained by the British laws, which are founded on the basis of humanity and public benefit, that the bodies of those who during life violated one of the most sacred rights of mankind, should after execution be devoted to a purpose which might make some atonement for their crime, by rendering their remains beneficial to that society which they offended.

In consequence of this regulation, I lately had an opportunity of performing some new experiments, the principal object of which was to ascertain what opinion ought to be formed of Galvanism as a mean of excitement in cases of asphyxia and suspended animation. The power which exists in the muscular fibre of animal bodies some time after all other signs of vitality have disappeared, had before been examined according to the illustrious Haller's doctrine of irritability; but it appeared to me that muscular action might be excited in a much more efficacious manner by the power of the Galvanic apparatus.

In performing these experiments, I had another object in view. Being favoured with the assistance and support of gentlemen eminently well skilled in the art of dissection, I proposed, when the body should be opened, to perform some new experiments which I never before attempted, and to confirm others which I had made above a year ago on the bodies of two robbers decapitated at Bologna.

DOI: 10.4324/9781003009337-12

To enlarge on the utility of such researches, or to point out the advantages which may result from them, is not my object at present. I shall here only observe, that as the bodies of valuable members of society are often found under similar circumstances, and with the same symptoms as those observed on executed criminals, by subjecting the latter to proper experiments, some speedier and more efficacious means than any hitherto known, of giving relief in such cases, may, perhaps, be discovered. In a commercial and maritime country like Britain, where so many persons, in consequence of their occupations at sea, on canals, rivers, and in mines, are exposed to drowning, suffocation, and other accidents, this object is of the utmost importance in a public view, and is intitled to every encouragement.

Forster, on whose body these experiments were performed, was twenty-six years of age, seemed to have been of a strong, vigorous constitution, and was executed at Newgate on the 17th of January, 1803. The body was exposed for a whole hour in a temperature two degrees below the freezing point of Fahrenheit's thermometer; at the end of which long interval it was conveyed to a house not far distant, and, in pursuance of the sentence, was delivered to the College of Surgeons. Mr. Keate, master of that respectable society, having been so kind as to place it under my direction, I readily embraced that opportunity of subjecting it to the Galvanic stimulus, which had never before been tried on persons put to death in a similar manner: and the result of my experiments I now take the liberty of submitting to the public.

Before I conclude this short introduction, I consider it as my duty to acknowledge my obligations to Mr. CARPUE, lecturer on anatomy, and Mr. HUTCHINS, a medical pupil, for the assistance they afforded me in the dissection. I was also much indebted to Mr. CUTHBERTSON, an eminent mathematical instrument maker, who directed and arranged the Galvanic apparatus. Encouraged by the aid of these gentlemen, and the polite attention of Mr. KEATE, I attempted a series of experiments, of which the following is a brief account.

EXPERIMENT I

ONE arc being applied to the mouth, and another to the ear, wetted with a solution of muriate of soda (common salt), Galvanism was communicated by means of three troughs combined together, each of which contained forty plates of zinc, and as many of copper. On the first application of the arcs the jaw began to quiver, the adjoining muscles were horribly contorted, and the left eye actually opened.

EXPERIMENT II

ON applying the arc to both ears, a motion of the head was manifested, and a convulsive action of all the muscles of the face: the lips and eyelids were also evidently affected; but the action seemed much increased by making one extremity of the arc to communicate with the nostrils, the other continuing in one ear.

EXPERIMENT III

The conductors being applied to the ear, and to the rectum, excited in the muscles contractions much stronger than in the preceding experiments. The action even of those muscles furthest distant from the points of contact with the arc was so much increased as almost to give an appearance of re-animation.

EXPERIMENT IV

In this state, wishing to try the power of ordinary stimulants, I applied *volatile alkali* to the nostrils and to the mouth, but without the least sensible action: on applying Galvanism great action was constantly produced. I then administered the Galvanic stimulus and volatile alkali together; the convulsions appeared to be much increased by this combination, and extended from the muscles of the head, face, and neck, as far as the *deltoid.* The effect in this case surpassed our most sanguine expectations, and vitality might, perhaps, have been restored, if many circumstances had not rendered it impossible.

EXPERIMENT V

I next extended the arc from one ear to the *biceps flexor cubiti*, the fibres of which had been laid bare by dissection. This produced violent convulsions of all the muscles of the arm, and especially in the *biceps* and the *coraco brachialis*, even without the intervention of salt water.

EXPERIMENT VI

An incision having been made in the wrist, among the small *filaments of* the *nerves* and *cellular membrane*, on bringing the arc into contact with this part, a very strong action of the muscles of the fore-arm and hand was immediately perceived. In this, as in the last experiment, the animal moisture was sufficient to conduct the Galvanic stimulus without the intervention of salt water.

EXPERIMENT VII

The short muscles of the thumb were dissected, and submitted to the action of the Galvanic apparatus, which induced a forcible effort to clench the hand.

EXPERIMENT VIII

The effects of Galvanism in this experiment were compared with those of other stimulants. For this purpose, the point of the scalpel was applied to the fibres, and even introduced into the substance of the *biceps flexor cubiti* without producing the slightest motion. The same result was obtained from the use of *caustic volatile alkali* and *concentrated sulphuric acid.* The latter even corroded the muscle, without inducing it to action.

EXPERIMENT IX

Having opened the *thorax* and the *pericardium*, exposing the heart *in situ*, I endeavoured to excite action in the *ventricles*, but without success. The arc was first applied upon the surface, then in the substance of the fibres, to the *carneæ columnæ*, to the *septum ventriculorum*, and lastly, in the course of the nerves by the *coronary arteries*, even with salt water interposed, but without the slightest visible action being induced.

EXPERIMENT X

In this experiment the arc was conveyed to the *right auricle*, and produced a considerable contraction, without the intervention of salt water, but especially in that part called the *appendix auricularis:* in the left auricle scarcely any action was exhibited.

EXPERIMENT XI

Conductors being applied from the spinal marrow to the fibres of the *biceps flexor cubiti*, the *gluteus maximus*, and the *gastrocnemius*, separately, no considerable action in the muscles of the arm and leg was produced.

EXPERIMENT XII

The *sciatic nerve* being exposed between the *great trochanter of the femur* and the *tuberosity of the ischium*, and the arc being established from the *spinal marrow* to the *nerve* divested of its *theca*, we observed, to our astonishment, that no contraction whatever ensued in the muscles, although salt water was used at both extremities of the arc. But the conductor being made to communicate with the *fibres* of the *muscles* and the *cellular membrane*, as strong an action as before was manifested.

EXPERIMENT XIII

By making the arc to communicate with the *sciatic nerve* and the *gastrocnemius* muscle, a very feeble action was produced in the latter.

EXPERIMENT XIV

Conductors being applied from the *sciatic* to the *personæal nerve*, scarcely any motion was excited in the muscles.

EXPERIMENT XV

THE *sciatic nerve* being divided about the middle of the thigh, on applying the conductors from the *biceps flexor cruris* to the *gastrocnemius*, there ensued a powerful contraction of both. I must here observe that the muscles continued excitable for seven hours and a half after the execution. The troughs were frequently renewed, yet towards the close they were very much exhausted. No doubt, with a stronger apparatus we might have observed muscular action much longer; for, after the experiments had been continued for three or four hours, the power of a single trough was not sufficient to excite the action of the muscles: the assistance of a more powerful apparatus was required. This shows that such a long series of experiments could not have been performed by the simple application of metallic coatings. I am of opinion that, in general, these coatings, invented in the first instance by Galvani, are passive. They serve merely to conduct the fluid pre-existent in the animal system; whereas, with the Galvanic batteries of Volta, the muscles are excited to action by the influence of the apparatus itself.

FROM the above experiments there is reason to conclude:

I

That Galvanism, considered by itself, exerts a considerable power over the nervous and muscular systems, and operates universally on the whole of the animal œconomy.

II

That the power of Galvanism, as a stimulant, is stronger than any mechanical action whatever.

III

That the effects of Galvanism on the human frame differ from those produced by electricity communicated with common electrical machines.

IV

That Galvanism, whether administered by means of troughs, or piles, differs in its effects from those produced by the simple metallic coatings employed by Galvani.

V

That when the surfaces of the nerves and muscles are armed with metallic coatings, the influence of the Galvanic batteries is conveyed to a greater number of points, and acts with considerably more force in producing contractions of the muscular fibre.

VI

That the action of Galvanism on the heart is different from that on other muscles. For, when the heart is no longer susceptible of the Galvanic influence, the other muscles remain still excitable for a certain time. It is also remarkable that the action produced by Galvanism on the auricles is different from that produced on the ventricles of the heart, as is demonstrated in Experiment the tenth.

VII

That Galvanism affords very powerful means of resuscitation in cases of suspended animation under common circumstances. The remedies already adopted in asphyxia, drowning, &c. when combined with the influence of Galvanism, will produce much greater effect than either of them separately.

To conclude this subject, it may be acceptable to the reader to have a short but accurate account of the appearances exhibited on the dissection of the body, which was performed with the greatest care and precision by Mr. Carpue. – "The blood in the head was not extravasated, but several vessels were prodigiously swelled, and the lungs were entirely deprived of air; there was a great inflammation in the intestines, and the bladder was fully distended with urine. In general, upon viewing the body, it appeared that death had been immediately produced by a real suffocation."

It may be observed, if credit can be given to some loose reports, which hitherto it has not been in our power to substantiate, that after this man had been for some time suspended, means were employed with a view to abridge his sufferings, by accelerating his death.

From the preceding narrative it will be easily perceived that our object in applying the treatment here described was not to produce re-animation, but merely to obtain a practical knowledge how far Galvanism might be employed as an auxiliary to other means in attempts to revive persons under similar circumstances.

In cases when suspended animation has been produced by natural causes, it is found that the pulsations of the heart and arteries become totally imperceptible; therefore, when it is to be restored, it is necessary to re-establish the circulation throughout the whole system. But this cannot be done without re-establishing also the muscular powers which have been suspended, and to these the application of Galvanism gives new energy.

I am far from wishing to raise any objections against the administration of the other remedies which are already known, and which have long been used. I would only recommend Galvanism as the most powerful mean hitherto discovered of *assisting* and increasing the efficacy of every other stimulant.

Volatile alkali, as already observed, produced no effect whatever on the body when applied alone; but, being used conjointly with Galvanism, the power of the latter over the nervous and muscular system was greatly increased; nay, it is possible that volatile alkali, owing to its active powers alone, might convey the

Galvanic fluid to the brain with greater facility, by which means its action would become much more powerful in cases of suspended animation. The well known method of injecting atmospheric air ought not to be neglected; but here, likewise, in order that the lungs may be prepared for its reception, it would be proper previously to use Galvanism, to excite the muscular action, and to assist the whole animal system to resume its vital functions. Under this view, the experiments of which I have just given an account, may be of great public utility.

It is with heartfelt gratitude that I recall to mind the politeness and lively interest shown by the members of the College of Surgeons in the prosecution of these experiments. Mr. Keate, the master, in particular proposed to make comparative experiments on animals, in order to give support to the deductions resulting from those on the human body. Mr. Bliche observed that on similar occasions it would be proper to immerse the body in a warm salt bath, in order to ascertain how far that powerful and extended coating might promote the action of Galvanism on the whole surface of the body. Dr. Pearson recommended oxygen gas to be substituted instead of the atmospheric air blown into the lungs. It gives me great pleasure to have an opportunity of communicating these observations to the public, in justice to the eminent characters who suggested them, and as an inducement to physiologists not to overlook the minutest circumstance which may tend to improve experiments that promise so greatly to relieve the sufferings of mankind.

AN EXTRACT FROM *MR. NICHOLSON'S JOURNAL*, RESPECTING SOME LATE GALVANIC EXPERIMENTS, WHICH WILL BE GIVEN IN THE AUTHOR'S WORK ON THAT SUBJECT, NOW IN THE PRESS.[1]

"M. Aldini, Professor at the Institute of Bologna, and nephew of the celebrated Galvani, after having made his experiments at the National Institute of France, has visited London, and given an accurate account of his experiments and discoveries to the Royal Society, before whom the same was read on the 25th last. I have the pleasure to communicate some of the principal facts which he has had the goodness to communicate to me, which appear calculated to throw much light on some of the most difficult phænomena of nature.

"Various philosophers have considered the metals as not absolutely necessary for the production of Galvanism, and Mr. Davy has proved it in the pile: It has also been indistinctly apprehended or conjectured in the way of theory, that the Galvanic or electric matter was excited, collected, or generated in the bodies of animals, where it was considered as the great cause or instrument of muscular motion, sensation, and other effects highly interesting, but very little understood. Professor Aldini has the distinguished merit of having placed this proposition in the rank of established truths. He has succeeded in exciting muscular contractions by the simple application of the nerves to the muscles of a prepared frog, without the least suspicion of any stimulus arising from contact. He has also given motion to the limbs of a small cold-blooded animal by the Galvanic energy of an animal with warm blood; an experiment never before imagined. He takes the head of an ox recently cut off, and applying the finger of one hand, wetted with salt water, to the spinal marrow, he holds in the other hand the muscle of a frog prepared (that is by dissection) in such a direction that its crural nerves shall touch the cervical muscles on the tongue of the ox. Every time of this contact strong contractions are produced in the frog. If a chain of several persons be formed holding hands, the same effect takes place; but the contacts do not produce any effect if the chain of connection be broken or interrupted. Here then we have the most decided substitution of the organized animal system in the place of the metallic pile: it is an animal pile; and the production of the Galvanic fluid, or electricity, by the direct or independent energy of life in animals, can no longer be doubted.

"The Professor has lately repeated the series of these experiments at Oxford, and showed, in the presence of Doctors Pegg and Bancroft, that the nerves of a prepared frog, disposed in the manner here stated, approach very sensibly to the muscles of the warm-blooded animal, and exhibit a real attraction hitherto unknown in natural philosophy and physiology. He invites philosophers to vary and repeat this phænomenon, which has been confirmed by different philosophers, particularly by the celebrated Felice Fontana of Florence. Galvanism, by these facts, is shown to be animal electricity, not merely passive, but most probably performing the most important functions in the animal œconomy. This power appears not to be confined in its operation to the motion of the muscles, but also appears to be of importance in the secretions. Aldini has given strength to this conjecture, by subjecting the urine to the power of the artificial Galvanic stream, and he found it productive of a separation of the principal combined parts of that fluid, which were considered as of much importance by the celebrated Professors Senebier and Jurine, of Geneva.

"A very ample series of experiments were made by Professor Aldini, which show the eminent and superior power of Galvanism beyond any other stimulant in nature. In the months of January and February last, he had the courage to apply it at Bologna to the bodies of various criminals who had suffered death at that place, and by means of the pile he excited the remaining vital forces in a most astonishing manner. This stimulus produced the most horrible contortions and grimaces by the motions of the muscles of the head and face; and an hour and a quarter after death, the arm of one of the decollated bodies was elevated eight inches from the table on which it was supported, and this even when a considerable weight was placed in the hand. These experiments have since been confirmed in various parts of Italy, particularly at Turin, by Professors Giulio, Vassali, and Rossi.

"These are not experiments of pure curiosity, but offer very encouraging prospects for the benefit of mankind, in disorders of the head, and in apoplexies. Professor Aldini means to apply part of his time in London, in communicating these important subjects of information to physicians, as he has already done in Paris, in which place he made some applications of his discoveries, chiefly at the Hospital de Salpetriere, in company with Doctor Pinel[*]. The application of Galvanism in melancholic insanity is absolutely new, and of great interest. He perfectly cured two patients at Bologna of this disorder; and on that account he is more desirous of recommending the trial in an affliction so distressing, against which the present system of physic has so little to offer.

"It appears to be equally promising in cases of apoplexy. Aldini thinks it may be highly useful in recovering drowned persons, and on that account he is desirous of communicating with the Society established at London for the recovery of these unfortunate individuals. An experiment lately made at Paris adds much to his expectations. At the Hospital de la Charité he showed the pupils that Galvanism applied to the trunk of a dog, to the spinal marrow, and the intestines, caused the lungs to act in such a manner, that the air that issued from the *aspera arteria* twice in succession extinguished a large candle placed opposite. Now, as in most cases little more is required in drowned subjects than to set the organs of respiration into action, it is to be presumed that the action of Galvanism may be of the highest utility in these cases. Many precautions are necessary to be used in the administration of this powerful remedy in lunacy or apoplexy, which will be detailed in a large work which Professor Aldini intends to publish in this country before his return to Italy. In the mean time the reader will, doubtless, receive satisfaction from this short notice he has enabled me to give of his labours, on a subject which promises greatly to extend the limits of natural science, and may be reasonably expected to add to the powers which man is enabled to exert for his own benefit over the numerous beings around him."

Note

1 Whose *Treatise sur la Manie* is well known in this country.

Editorial Headnote

John Aldini, *An Account of the Late Improvements in Galvanism* (London: Cuthell and Martin, 1803), 101–7, 110–12.

8

AN ACCOUNT OF THE LATE IMPROVEMENTS IN GALVANISM

John Aldini

Source: London: Cuthell and Martin, 1803, 101–107, 110–112

APPLICATION OF GALVANISM TO THE ORGANS OF HEARING AND OF SIGHT

By applying Galvanism to different parts of the face, a flash of light is excited in the eyes, which is stronger or weaker according to the nature of the parts to which it is directed. These organs, though delicate, are always affected in such a manner that the mechanism of the eye sustains no injury from the metallic arcs when they are made to communicate with only two plates of different metals. I received an account of this phænomenon at Milan, several years ago, from the celebrated Volta, who produced it by applying a conductor of zinc in such a manner, that one end of it touched the bulb of the eye, and the other the tip of the tongue armed with a plate of tin. In administering Galvanism for diseases of the eyes, it is much better to employ the pile. To excite the appearance of a flash of light, it is not necessary that the eyes should be open; it takes place even if they are shut, and covered with a bandage in a darkened apartment. Some of the partisans of Euler, perhaps, may here maintain, that this phænomenon of Galvanism is a deception, and that no light can be really excited in such cases, as the production of light depends on the emanation of rays from a luminous body which penetrate into the interior part of the eye. But without entering into the examination of this question, which is foreign to my principal object, I shall only observe, that long before the theory of Galvanism was known, the celebrated Darwin had found that when the eyes are shut, there are certain internal stimulants which are capable of producing the appearance of light and of colours.

In no case is the difference between Galvanism and electricity more apparent than in the application of the former to the sight. Very little benefit had been obtained by means of a metallic point inclosed in a glass tube, which directed the electric charge from a jar to the cornea; and besides, the shock of a Leyden jar applied directly to so delicate an organ as the eye could not but be attended with danger. Galvanism applied to the exterior part of the eye, in the same place where

DOI: 10.4324/9781003009337-13

a stream of electricity would have had no influence whatever, has always produced a certain effect on the organ of sight, as is proved by the following experiments.

EXPERIMENT I

If you touch with one hand the bottom of the pile, and at the same time apply to the summit different parts of the face moistened with salt water, a flash of light will be excited in the eyes. The same result will be obtained, if instead of touching the bottom of the pile with the hand you touch it with the sole of your foot. No flash of light is observed when the Leyden flask is employed in the same manner.

EXPERIMENT II

Having observed the preceding phænomenon in myself, and excited it in others, I was desirous of proving it in regard to several persons at the same time during the course of my public lectures, and for that purpose I made use of the following apparatus. I arranged two metallic plates in a horizontal position, at the distance of nine inches from each other, so that six persons with their hands dipped in salt water could touch the lower plate, and the upper one with the tips of their tongue. A charged Leyden flask placed between the two plates being then discharged, the whole of the persons experienced a violent shock, but perceived no flash of light. It is well known however that, when a similar arc is formed with the interposition of a very strong pile, a flash of light is constantly observed; though the force in the second case is much less than that excited by the explosion of the Leyden flask. The same result will be obtained by bringing the upper plate into contact with the nose.

EXPERIMENT III

As my pupils took much interest in this research, some of them suspected that the light of the apartment might perhaps have effaced that excited by the electricity. I therefore made the apartment entirely dark; and one of them taking a Leyden flask applied it to the point of the nose of another person with whom he was in communication, by laying hold of him with the other hand. By these means a very strong shock was given, but no flash of light was observed. This experiment was repeated, making the person who received the shock of the Leyden flask to remain some time before in the dark, that his eyes might be better enabled to perceive any faint light that might be excited: but the result was still the same. To those who refer Galvanism to the common laws of electricity, it will be difficult to comprehend the cause of the different action exercised by the latter on the organ of sight. But as it is not my intention at present to enter into any discussion on this subject, I shall leave it to philosophers to assign a reason for this phænomenon, and only observe, that the properties above indicated will be sufficient to authorize medical practitioners to prefer, in certain cases, the administration of Galvanism by the pile to that of common electricity.

Before I proceed directly to the medical administration of Galvanism to the organ of sight, I think it necessary to distinguish four classes of blind persons whose cases ought to be considered separately.

The first belongs to those who from their birth have been deprived of the valuable blessing of sight.

The second comprehends those become blind in consequence of some great læsion, or some derangement in the solids or in the fluids which constitute the mechanism of the eye.

The third, those who have become blind by some morbid action, though the mechanism of the eye has been little affected, and though no impediment has occurred but in regard to the action of the optic nerve.

The fourth class comprehends those who, though not actually deprived of sight, have it much weakened in consequence of disease, or of some other cause.

The administration of Galvanism does not hold forth much hope of a cure to persons belonging to the first two classes. I however resolved to attempt some experiments on this subject at Bologna; but though there were a great many blind in that city, I found that they had become so by the malignant influence of the small pox. This observation will, I hope, be of service to the pursuits of the celebrated Dr. Jenner, and of all those who exert themselves to promote the beneficial practice of vaccine inoculation.

Being deprived, at Bologna, of any opportunity of trying the effects of Galvanism in cases of persons born blind, I galvanised several who had lost their sight at a very early age. I first applied the Galvanism to the arms of five blind persons, some of whom had lost their sight thirty, and others forty years and even more. By this method they were familiarised with the idea of its mechanical action, and learned to distinguish it from every other sensation. I then applied Galvanism to the lips, and to the tip of the nose, in a darkened apartment; but in three instances only the patients had a real perception of light, to which they had been so long strangers. I then applied it to cases of amaurosis, and at first had confident expectation of effecting a perfect cure. One of my patients was a woman, whom this disorder had deprived of the sight of one eye, while that of the other was much weakened. After administering the Galvanism different ways, I observed that the eye totally blind began to have a perception of light, and that the sight of the one which was weakened became much stronger. I then took a book, which I held at a considerable distance from her; and removing it gradually further as the Galvanism was administered, I observed, in consequence of this method, that the patient's sight daily improved. But I must freely confess that the success obtained, though at first flattering, was not of long duration; and that when the Galvanism was discontinued a great deal of the benefit which had been obtained was again lost. On this account, I was discouraged from administering Galvanism any more in such cases. I am however of opinion, that by varying the method of administration it may be attended with some utility. I shall here observe, that having once had occasion to administer common electricity in a similar case of amaurosis, I was never able to excite the perception of a flash in the eyes of the patient, though the electricity was applied directly to the eye itself.

. . .

. . .

APPLICATION OF GALVANISM IN CASES OF ASPHYXIA AND DROWNING

I mentioned in the second part of this work the great influence which Galvanism has in cases of asphyxia, and the preference which ought to be given to it in comparison of other stimulants. Though the observations offered in that part are sufficient to prove my proposition, I shall add to them the following experiments:

EXPERIMENT I

Some dogs and cats were immersed in a large pond till they gave no external signs of respiration, or of muscular motion; and Galvanism being immediately administered to them, according to the methods already described, they were sometimes restored to life. I make use of the term ‘sometimes,’ because, if animals are immersed in water for a longer period than their organization can bear, and if the vital powers are really destroyed, it is evident that it will be impossible to restore them to life by any physical process whatever. I obtained the same results from to animals thrown into a state of asphyxia in different ways.

EXPERIMENT II

Having applied Galvanism to the trunk of a dog, in the *Hôpital de la Charité* at Paris, air seemed to escape from the tracheal artery on every application of the arc. Being requested to repeat and confirm this interesting experiment, I found myself under the necessity of sacrificing a new victim to my Galvanic researches. As it was necessary to examine the phænomenon while the body was in that state of vitality most proper for the observation, I exposed the trunk of another dog recently killed to the Galvanic action; and having placed a taper near to the tracheal artery, it was extinguished twice in succession by two applications of Galvanism. By repeating this experiment, in Mr. Wilson’s anatomical theatre, Great Windmill-street, and in the theatres of Guy’s and St. Thomas’s Hospital, London, I found that the taper could be extinguished a greater number of times.

These experiments give me sufficient reason to hope that Galvanism may be administered with some advantage in cases of drowning. But as I never had an opportunity of trying the effects of this stimulant in such cases, I have requested several medical practitioners to pay attention to this application of Galvanism, which may be of the utmost importance to the cause of humanity. I have already mentioned that the method which I propose is exceedingly simple; that no anatomical operation whatever is required; that it is attended with no danger; and that the possibility of saving the life is in every case respected. Nothing is necessary but to immerse in salt water one of the hands of the person subjected to the operation, and then to apply the Galvanic current to one of the ears and to the surface of the salt water.

Dr. Lettsom, a very zealous member of the Humane Society, having been present at some of my experiments, I requested that he would recommend the application of Galvanism in cases of drowning. He assured me that he would use his endeavours to cause my method to be tried, and I had several conferences with him on the subject, that I might communicate to him such information as might tend to facilitate the application of it. As the Galvanism in such cases ought to be administered with great promptitude, we agreed that the apparatus of the trough is preferable to that of the pile; and we contrived the plan of a portable box to contain a trough, two arcs, and a solution of common salt. Such an apparatus will be exceedingly convenient, and may easily be employed in all cases of drowning and of asphyxia.

Part 4

CHEMISTRY/PHYSICS

Editorial Headnote

Claude Louis Berthollet, *An Essay on Chemical Statics* (London: J. Mawman, 1804), I, vii–xiv

Claude Louis Berthollet, 1748–1822, pioneer of modern chemistry in France, here emphasising the importance of direct observation.

9

AN ESSAY ON CHEMICAL STATICS

Claude Louis Berthollet

Source: London: J. Mawman, 1804, I, vii–xiv

THE powers which produce chemical phenomena are all derived from the mutual attraction of the moleculæ of bodies, to which the name of affinity is given, to distinguish it from astronomical attraction.

It is probable that they are the same property; but astronomical attraction being only exerted between masses placed at a distance, in which the figure of the moleculæ, their intervals, and peculiar habitudes, have no influence, its effects, in all cases proportionate to the mass, and in the inverse ratio of the square of the distances, may be the object of accurate calculation: the effects of chemical attraction or affinity are, on the contrary, so changed by particular and frequently indeterminate conditions, that they cannot be deduced from any general principle; but must be considered in succession. There are but few of these effects which can be sufficiently disengaged from all the other phenomena to be rendered capable of precise calculation.

It is, therefore, by observation alone that the chemical properties of bodies, or those affinities by means of which they exert a reciprocal action in a determinate situation, can be estimated; nevertheless, since it is very probable that affinity differs only in its origin from the general attraction, it ought to be also submitted to those laws which mechanics has established for the phenomena arising from the action of the mass, and it is natural to suppose that the more of generality there is in the principles from which the chemical theory is derived, the greater must be their analogy with those of mechanics; but it is only by means of observation that they can attain to that degree which may be already foreseen.

The immediate effect of the affinity which a body exerts is always a combination; here all the effects produced by chemical action are the consequence of the formation of some combination.

Every substance which has a tendency to enter into combination, acts in the ratio of its affinity and of its quantity. These facts are the ultimate end of every chemical observation.

But, 1st. The different tendencies to combination may be considered as so many forces which contribute to a result, or which partly destroy it by their opposition;

DOI: 10.4324/9781003009337-15

these forces must, therefore, be distinguished in order to obtain an explanation of the phenomena which they produce, or to compare them together.

2d. The chemical action of a body does not depend solely upon the affinity peculiar to its component parts and upon the quantity; it also depends upon the state in which those parts are found, whether it be that of actual combination, which causes a larger or smaller portion of their affinity to disappear, or that, by their dilatation or condensation, their reciprocal distances are varied: it is these conditions which, by modifying the properties of the elementary parts of a substance, form that which I call its constitution: to obtain the analysis of chemical action, not only each of these conditions must be appreciated, but also every circumstance with which they have any connection.

The properties of bodies which can thus modify affinity have also other effects, independent of those which produce the combination, and which are the subjects of the different parts of physics. There are even many phenomena which, although they may be produced, wholly or in part, by affinity, ought nevertheless to be considered in another view, either because affinity contributes too slightly to their production, or because experiment has not hitherto been able to determine the particular affinities from which they arise. All those are called physical properties which do not seem to depend immediately on affinity.

Hence it follows, that a connection frequently exists between the physical properties and the chemical properties; that in many cases recourse must be had to both of them for the explanation of a phenomenon to which they may each contribute, and that it is necessary to establish an intimate relation between the different sciences of physics, that they may mutually elucidate each other.

The principles established on the results of facts observed under every point of view, and the explanation of chemical phenomena, founded on their relations to all the properties of which they are the consequences, constitute Theory, which should be distinguished into general theory and partial theories.

Some sciences may reach a certain degree of perfection without the aid of any theory, solely by means of an arbitrary order, established from the observations of those natural facts which are principally the subject of them; but this is not the case with chemistry, in which the observations ought, in almost every case, to arise from experiment alone, and in which the facts result from the factitious union of the circumstances from which they must be produced. In making an experiment it is necessary to have some object in view, and to be guided by an hypothesis; and to obtain any advantage from the observations, they must be compared under certain conditions, and, at least, some of the circumstances necessary to the production of every phenomenon observed, must be ascertained, to the end that they may be reproduced. Thus, suppositions more or less illusory, and even those chimeras which at present appear ridiculous, but which have employed the most laborious investigations, were necessary to the infancy of chemistry: by their means facts were multiplied; a great number of properties were ascertained, and many arts were brought to perfection.

Still Chemistry was only enlarged with incomplete remarks and partial theories, unconnected with each other, which arose in succession like the caprices of the imagination, and had no relation to any general law: arrogant and insulated from every other species of knowledge, the greater the number of acquisitions made to it, the further was it removed from the character of true science.

It is only since the period that affinity has been recognized as the cause of all combinations, that Chemistry could be regarded as a science which began to have general principles: from that time its object has been to bring the succession of combinations, which different elements are capable of forming, into a regular system, and to ascertain the proportions which enter into these combinations.

Bergman extended the application of this first principle very considerably: he pointed out the greatest number of the causes which might disguise or produce variations in its effects: on it he founded the processes of different chemical analysis, which brought it to a degree of precision unknown before his time.

A great number of phenomena, however, are dependent on the combination of oxigen, which, of all substances, is that whose affinities appear to be the most active, and its existence was not even known: hypotheses were, therefore, substituted for the action which it exercises. Priestley had no sooner made known this substance, which performs so important a part, than Lavoisier ascertained its combinations, and ascribed the numerous effects produced by it to the true cause. The great light spread by his immortal discoveries, not only over those phenomena dependent on them, but also over the action of many other gases discovered at the same period, obtained, for the revolution which it produced, the honour of being regarded as a new and general theory.

The accurate investigation of a cause, equally powerful from the modifications produced by it in the results of affinity, that of the action of heat, was also necessary to explain the greater number of phenomena: it is to Black that the discovery of the fundamental properties of heat are due; since his time they have engaged the attention of many philosophers; but they were reduced to well-defined laws in an intelligent Memoir, for which we are indebted to Laplace and Lavoisier.

It is therefore evident that Chemistry has acquired, in our time, the knowledge of those generative properties which accompany all chemical action, and which are the source of all the phenomena produced by it: this science has therefore been founded upon principles which have been rapidly applied to every branch of knowledge within its compass.

Editorial Headnote

John Dalton, *A New System of Chemical Philosophy* (Manchester: Bickerstaff, 1810)

John Dalton, 1766–1844, famed principally for atomic theory and for furthering the understanding of colour blindness, was a famed chemist and physicist. Here he dwells on the labours of experimental science and his commitment to rely only upon his own experience.

10

A NEW SYSTEM OF CHEMICAL PHILOSOPHY

John Dalton

Source: Manchester: Bickerstaff, 1810

When the first part of this work was published, I expected to complete it in little more than a year; now two years and a half have elapsed, and it is yet in a state of imperfection. The reason of it is, the great range of experiments which I have found necessary to take. Having been in my progress so often misled, by taking for granted the results of others, I have determined to write as little as possible but what I can attest by my own experience. On this account, the following work will be found to contain more original facts and experiments, than any other of its size, on the elementary principles of chemistry. I do not mean to say that I have copied the minutes of my note-book; this would be almost as reprehensible as writing without any experience; those who are conversant in practical chemistry, know that not more than one new experiment in five is fit to be reported to the public; the rest are found, upon due reflection, to be some way or other defective, and are useful only as they shew the sources of error, and the means of avoiding it.

Finding that my design could not be completed, without a second volume, I was desirous to finish the 5th chapter, which treats of the compounds of two elements, in the part now edited; but the work is enlarged so much, and the time is so far advanced, that I have been obliged to omit two or three important sections, particularly the metallic oxides and sulphurets, which I am aware will demand no inconsiderable share of attention. After these are disposed of, the 6th chapter will treat of compounds of 3 or more elements; this will comprehend the vegetable and other acids not yet noticed, the hydrosulphurets, the neutral salts, compound combustibles, &c. &c.

Whatever may be the result of my plan to render the work somewhat like complete, by the addition of another volume, I feel great present satisfaction in having been enabled thus far to develope that theory of chemical synthesis, which, the longer I contemplate, the more I am convinced of its truth. Enough is already done to enable any one to form a judgment of it. The facts and observations yet in reserve, are only of the same kind as those already advanced; if the latter are not sufficient to convince, the addition of the former will be but of little avail. In the mean time, those who, with me, adopt the system, will, I have no doubt, find it a very useful guide in the prosecution of all chemical investigations.

DOI: 10.4324/9781003009337-16

In the arrangement of the articles treated of, I have endeavoured to preserve order; namely, to take such bodies as are simple, according to our present knowledge; and next, those bodies that are compounds of two elements; but in this I have not always succeeded. For, in some instances, it has not been quite clear what was simple, and what compound; in others, the compounds of three or more elements have been so intimately connected with those of two, that it was found impracticable to give a satisfactory account of the latter, without entering more or less into a description of the former.

In regard to nomenclature, I have generally adopted what was most current; perhaps, in a few instances, my peculiar views may have led me to deviate from this rule. I have called those salts *carbonates*, which are constituted of one atom of carbonic acid united to one of base; and the like for other salts. But some moderns call the *neutral* salts carbonates, and the former *subcarbonates;* whereas, I should call the neutral carbonates of soda and potash *supercarbonates*, consisting of two atoms of acid and one of base. I have, however, continued to call the common *nitrates* by that name, though most of them must be considered on my system as *supernitrates*. I am not very anxious upon this head, as it is evident that if the system I proceed upon be adopted, a general reformation of nomenclature will be the consequence, having a reference to the *number of atoms*, as well as to the *kind of elements*, constituting the different compound bodies.

Nov. 1810.

Editorial Headnote

Michael Faraday, *Chemical Manipulation; being instructions to students in chemistry, on the methods of performing experiments of demonstration or of research, with accuracy and success* (London: Phillips, 1827), i–ix

Michael Faraday, 1791–1867, known principally for research on electromagnetism and electrochemistry, was appointed Fullerian Professor of Chemistry at the Royal Institution of Great Britain in 1833. With a background of modest means and little formal education, here we find him extolling the virtues of experimental method with the express purpose of creating 'new facts'.

11

CHEMICAL MANIPULATION; BEING INSTRUCTIONS TO STUDENTS IN CHEMISTRY, ON THE METHODS OF PERFORMING EXPERIMENTS OF DEMONSTRATION OR OF RESEARCH, WITH ACCURACY AND SUCCESS

Michael Faraday

Source: London: Phillips, 1827, i–ix

CHEMISTRY is necessarily an experimental science: for as facts are the data from which its conclusions are drawn, and the evidence by which its principles are supported, a constant appeal to them is necessary; and yet so small, comparatively, is the number of these necessary elements presented to us spontaneously by nature, that were we to bound our knowledge by them, it would extend but a very small distance indeed, and that even in an uncertain manner. To supply the deficiency, new facts have been created by *experiment*, the contrivance and hand of the Philosopher having been employed in their production and variation. In reference to the varieties of inert matter, all their possible forms, states, and properties, and the powers which influence them, Chemistry, if occupied only in the observation of such phenomena as are presented by nature, would do little more than record a state of things approaching to quiescence, which has resulted from the active exertion of the inherent powers of matter; and the Chemist would have but little opportunity of observing substances in their energetic state, or of witnessing the actual exertion of the powers inherent in them. Even when such appearances might naturally be presented, either their vastness, their complication, or their rarity, would in many cases prevent the deduction of correct conclusions.

Hence the importance of multiplying facts by every means in our power whilst engaged in the pursuit of this science. If it were our object to learn as much as possible

DOI: 10.4324/9781003009337-17

of the nature of a substance taken at random from the surface of the earth, how slight would be the amount of information derived from a consideration of it in its natural state;—we should be able to decide perhaps that the air did not affect it, that light did not sensibly injure it, that the rain or dews did not dissolve it, that the difference between summer and winter did not apparently alter it, that the ground beneath was not affected by it, and perhaps some other points respecting its nature might be ascertained: but what are these compared to the knowledge gained by submitting it to a high temperature, either alone or in contact with other substances; by subjecting it to the action of other bodies as solvents under peculiar circumstances, and operating upon it in the numberless ways which art and experience may dictate to us. Unaltered as it appeared before, it may now change its form, yield new substances and enter into new combinations, and instead of being the inert lump it appeared, may prove an active and powerful agent in many of the purposes of civilized life.

Such then are the advantages which result from experiment; nor is the case at all exaggerated, for in Chemistry it may safely be stated that more than nine-tenths of the facts upon which the science is founded are thus evolved by artificial means. Without indeed the great body of truth thus furnished, the science could not have existed; for though it is not exclusively, it is preeminently experimental, being in this respect strikingly distinguished from Astronomy, Botany, Zoology, and other sciences, which have regard to appearances and phenomena presented by nature. Hence it derives many of its peculiar charms.

These considerations are abundantly sufficient to shew the intrinsic importance of experiments; and when to them are added those which arise from the great extent and dominion of the science itself over the powers and properties of all matter, and its influence in administering to the wants, comforts and pleasures of life, little need be urged in extenuation of an attempt to facilitate the acquirement of the art of making experiments by students in Chemistry.

Experiment has two principal objects; the extension of our present knowledge, and the proof or demonstration of knowledge previously acquired; in both of which it is essentially necessary to the progress of chemical discovery. There are also two parts in an experiment; first, it has to be devised; its general nature and principles are to be arranged in the mind, and the causes to be brought into action, with the effect to be expected, properly considered; and then it has to be performed. The ultimate objects of an experiment, and also the particular contrivance or mode by which those ultimate results are to be produced, being mental, there remains the mere performance of it, which may properly enough be expressed by the term *manipulation*.

Notwithstanding this subordinate character of manipulation, it is yet of high importance in an experimental science, and particularly in Chemistry. The person who could devise only, without knowing how to perform, would not be able to extend his knowledge far, or make it useful; and where every doubt or question that arises in the mind is best answered by the result of an experiment, that which enables the philosopher to perform the experiment in the simplest, quickest, and most correct manner, cannot but be esteemed by him as of the utmost value. It is indeed to him like the external senses to the mind, a channel of information by which things before unperceived are made known, and for which he has continual use.

There are many experiments, and even whole trains of research, which are essentially dependent for success on mere manipulation; such for instance are various analytical processes, in which the principles of the process and the modes of detecting and separating the substances being well known, the accuracy with which they are successively separated, and their quantities correctly ascertained, depends entirely upon manipulation. Such is the case in most of the analyses of ordinary siliceous, calcareous, or aluminous stones: such is also the case in the analysis of organic substances by any of the processes recommended in chemical works: it is the same with the manufacture of many chemical preparations; and the separation of four or five gases, a problem of frequent occurrence, is to the chemist who has made but moderate progress even in his studies, a matter entirely of manipulation.

By accurate and ready manipulation, therefore, an advantage is gained independent of that belonging to the knowledge of the principles of the science, and this is so considerable, that of two persons having otherwise equal talents and information, the one who manipulates best will very soon be in advance of the other; for the one may obtain satisfactory results from his experimental enquiries, while the other is left in doubt or led astray by his imperfect reasonings. This advantage may be illustrated by the use of the tinder syringe, a small instrument consisting of a cylinder about half an inch in diameter, and three or four inches in length, closed at one end and fitted with a piston, to the extremity of which a piece of amadou is fastened: by forcing the piston down and compressing the air *suddenly*, so much heat is evolved as to fire the tinder. Some persons cannot perform this simple experiment, whatever may be the strength or alertness which they endeavour to bring into action, whilst others with a very slight force and the mere approach of their hands towards each other in the air, will in every instance obtain the effect desired and produce the required ignition. Were this a new experiment to the persons making it, the object being by its institution to prove whether air when highly compressed gives out much heat or not, the first person would either come to a wrong conclusion, or, if he doubted the success of his experiment, would arrive at none at all, whilst the second would be enabled to form a correct and affirmative conclusion, and thus would have added an important fact to his previous knowledge.

In other cases where the appearances may be such that we have no method of anticipating, or when produced, of correcting them, as in the habitudes of an unknown substance or its action upon other bodies, then careful manipulation is of the utmost importance: without it the appearances produced may arise or be modified by extraneous substances present, or by other causes overlooked; and the conclusions may be erroneous at the time when it is most important they should be correct, because the subject is new, and because from its novelty but a small proportion of previous knowledge will bear upon the point and help us to correct the error. Nor is the tyro alone thus liable to be misguided; for it would not be difficult to point out instances where the most acute minds have in this way been led to false conclusions.

Another consequence of skilful manipulation is, that by its means a train of research may frequently be carried much farther than its first object, and in two ways. In the first place the occurrence of clear and distinct phenomena, besides proving satisfactorily the direct object of inquiry, frequently suggests to the mind collateral views which, pursued and extended, terminate in additional chains of information and discovery; whereas if the experiments be less clear, though sufficient may be distinguished to satisfy the mind on the subject in question, nothing more is done and no new object or view arises. In the second place the operations may be performed on comparatively very minute quantities, and that which in the hands of one person had hardly sufficed to supply very general information of its nature, may in those of another be made to yield matter for a full and minute investigation, terminating in the developement of its nature and habits, with a perfection equal to that obtained with the largest quantities. Frequent illustrations of the importance of manipulation in thus effecting all that can be desired with small quantities, is afforded in the occasions that arise juridically in testing for arsenic; for the fractions of a single grain, will in the hands of some persons afford the most striking and convincing proofs, whilst many grains will in the hands of others present no satisfactory conclusion.

When the substance under examination is rare, and that is frequently the case both in natural and artificial productions, the facility of working with small quantities is of the highest importance, as otherwise the opportunity of gaining information may be entirely lost, or if preserved is retained only at a great expense. There existed in the British Museum a small fragment of a black stone, the source and history of which was unknown; it was unique, no other specimen being in the Museum or known to be in existence; yet as it presented some peculiar characters, Mr. Hatchett was induced to examine it, and working with a portion of the stone weighing not more than 200 grains, he was enabled to discover in it a new metal, which he distinguished by its various characters from all those previously known, and which he named Columbium. Ekeberg afterwards discovered a metal which he named Tantalium, conceiving it to have been observed and distinguished for the first time by himself; but Dr. Wollaston, who examined it and compared it with Columbium, was able to identify it with that metal, although he had not more than five grains of the stone from the British Museum upon which to make his experiments.

Finally, habits of correct and delicate manipulation very much facilitate experimental enquiries at all times. It is not in difficult researches only that it is desirable, but even in such common operations as testing for lime, or iron, or sulphuric acid, its advantages become manifest; for either time is shortened, or the apparatus considered as necessary is diminished, or effectual substitution is made for those that may be wanting, and thus the experiment becomes easy, where otherwise it would be considered impossible. Besides facilitating such inquiries, it also diminishes the expense both in materials and apparatus, and it produces beneficial habits in the mind by exercising it both in invention and perception, even in this subordinate part of its operations. "Nothing," as Dr. Johnson observes, "is to be considered as a trifle by which the mind is inured to caution, foresight, and circumspection. The same skill, and often the same degree of skill, is exerted in great and little things."

The importance of instruction in manipulation has long been felt by the author during his professional experience as a public and private teacher of Chemistry in the Royal Institution; and the deficiency existing in the means of teaching it, induced him to think he might perform an acceptable service by putting together such information on the subject as there was reason to suppose would be generally useful to the student. No book contains those minute directions which are necessary in the present extensively cultivated state of the science, nor can verbal instruction teach that perfection of manipulation which is only to be gained by constant operation; but there is so much that can be taught, so much that can be suggested by such instruction, that it seems extraordinary that not one of the many treatises upon Chemistry has been devoted to this subject, especially when it is considered that of the great numbers who now desire, or are assumed to have some knowledge of Chemistry, very few have access to competent practical sources. Lavoisier's elements is the work which appears to the author to contain the best general directions, but every pupil to whom it has been shewn, has found it to fall far short of his necessities.

Such are the considerations which have given rise to the present work, and under their influence it has been composed. The object of the volume is to facilitate to the young chemist the acquirement of manipulation, and by consequence, his progress in the science itself. It does not attempt to inculcate the *principles* of the science, but the *practice*; neither does it claim to teach *a habit of reasoning*, but has solely in view the *art of experimenting*; and though sometimes it may be necessary to speak of the properties of bodies, or to draw conclusions, it will always be done in subordination to the main object. The volume indeed is not intended to supply the place, or imitate in its nature, any one of the numerous and useful works on Chemistry now extant, but to be rather an accompaniment to all of them, supplying that portion of knowledge which however essential to the learner, was not of a nature to consort with their more scientific contents.

In the pursuit of this object it is intended to describe

The conveniences and requisites of a laboratory.
Chemical apparatus and its uses.
The methods of performing chemical operations.
The facilities acquired by practice, and
The causes which make experiments fail or succeed.

Although a laboratory will be described in a complete state, well appointed, and with all the conveniences that the author is acquainted with, yet it is intended that the directions shall be such as to enable the experimenter to perform the operations, when desirable, with the smallest number of requisites; for though many may wish to be made acquainted with all that is useful, yet to far more a knowledge of the few essentials are of the greatest consequence. To omit the description of a complete laboratory in a work devoted to chemical experiment would of course be erroneous; but it would certainly be a much greater error to omit shewing, as far as the author's knowledge will serve him, how many things may be dispensed with in cases of necessity, and how few are the absolute requisites for the greater number of operations. The general principles upon which apparatus is formed will be inculcated for the purpose of leading the student to the *ready substitution* of one thing for another, and to that *contrivance*, by which the wants of the operator may be obviated; and therefore *small, temporary*, and *generally useful* apparatus, will be pointed out as often as possible.

It is not intended to describe processes particularly, such as the preparation of each of the acids, or of the alkalies, or of other important substances; or such as are numerous in the chemical arts, though reference will probably be frequently made to them. The work is intended principally to assist in obtaining a knowledge of the chemistry of research, and not the chemistry of the arts, or rather, not of the ultimate and refined processes of chemical preparations; otherwise than as, the principles being the same, the instruction which is advantageous in the one case will be useful in the other. The book is principally for beginners, but this will not induce the author to reject as improper any information he may have to give relative to the facilities of making experiments, though they may relate to the production of refined and abstruse results. Lastly, it may be remarked, that it is intended for a country where most of the requisites are supplied in trade.

Professedly disclaiming a scientific character, the arrangement of the volume is one of mere convenience, but it has not been adopted without considerable thought. As the work was not intended to teach the principles of Chemistry, but to be a useful laboratory companion, that arrangement was considered as best, which though not scientific, seems most convenient for including the different parts of manipulation, without any great violence to the relation of the whole subject. It has therefore been divided into *Sections*, or principal divisions; the object of each, however, has not always been strictly adhered to alone, since it has often been found convenient and useful to include information on other points, which nevertheless have always some degree of connexion with the subject under consideration. Those who may read the work for instruction will probably not find many difficulties in the arrangement, particularly as they will be aided by a copious index.

Such are the objects of the present book, and the means by which their attainment has been attempted. That it is faulty the author has no doubt; indeed he is thoroughly convinced it cannot be otherwise. It contains little more than the experimental practices and methods of one person; and though in the more ordinary operations of chemistry the author's may resemble those of most other chemists, and therefore be well adapted for general adoption, yet in those peculiar facilities that others may have attained by operating, it must necessarily be imperfect. Acquisitions of this kind made by any chemist must be in relation to his particular trains of research; and as there are but few who possess the power of ranging through the whole of this extensive science, it cannot be expected that every one should have perfect knowledge of the facilities belonging to all departments of experimental chemistry. The author has been anxious that omissions of importance should be as few as possible.

Finally, he has to observe with reference to the pupil's expectations in consulting this book, that even supposing it were perfect, it could still only point out the methods for him to *practise*, and thus, aided by the principles of the science detailed in other works, and the experimental directions in the present volume, he must perfect his knowledge by study and the labours of his own hands.

Editorial Headnote

Justus von Liebig, *Instructions for the Chemical Analysis of Organic Bodies* (Glasgow: Richard Griffith, 1839), 1–13

Justus Freiherr von Liebig, 1803–1873, known as a pioneer of organic chemistry and as a forerunner in laboratory-based teaching, at the University of Giessen. In this rudimentary advice that showcases modern scientific methods, the emphasis is on observation, care, accuracy and the value of experience.

12

INSTRUCTIONS FOR THE CHEMICAL ANALYSIS OF ORGANIC BODIES

Justus von Liebig

Source: Glasgow: Richard Griffith, 1839, 1–13

THE Analysis of organic substances has for its object, to determine the nature and quantity of the elements which compose them; and is one of the most important departments of analytical chemistry. The method employed by the earlier chemists to obtain a knowledge of the chemical composition of organic bodies had not the smallest resemblance to the organic analysis of the present day. They subjected these bodies to destructive distillation, and judged of the difference in their composition by the products thus obtained.

It is only within the last thirty years that this department of chemistry has been cultivated on scientific principles; and all the lately proposed methods differ from one another only in the way in which those principles are carried out.

The simplest method of ascertaining the component parts of an organic compound, would seem to be, to endeavour to obtain its elements in a separate form; but it is obvious that, if we can obtain, instead of the elements in the free state, compounds, of known composition, of those elements with others, we can determine their quantity with equal accuracy.

Most vegetable substances contain carbon, hydrogen, and oxygen; a small number contain, besides these elements, nitrogen. Of these four simple substances, nitrogen alone can be obtained in a state of purity from organic compounds; but if all the carbon be converted into carbonic acid, and all the hydrogen into water, we can then calculate, with the utmost precision, the quantity of carbon and hydrogen from that of the carbonic acid and water. Even if the elements of organic substances could be separated from them in a state of purity, we should be forced to give the preference, in analysis, to the indirect method now employed, on account of its superior accuracy.

The method, then, which we employ to procure an exact knowledge of the composition of an organic compound, consists in the conversion of a known weight of the substance into carbonic acid and water; and the success of the analysis

DOI: 10.4324/9781003009337-18

only depends on the apparatus employed, in so far as that apparatus must allow us to collect these products without loss, and to determine their weight. When the compound contains nitrogen, this element is collected in the separate state; and the oxygen is always ascertained indirectly.

Gay-Lussac, and Thénard, the first chemists who executed organic analyses, used chlorate of potash for the combustion of organic bodies. The substance to be analysed was mixed with it, the mixture formed into pellets, and introduced in small quantities into a red hot tube of glass, placed vertically. The gaseous matter disengaged by the combustion was, by means of a lateral tube, collected in a jar over mercury.

The whole gas was accurately measured, and, the corrections for barometer and thermometer being made, caustie potash was introduced into the jar. After all the carbonic acid was absorbed, there remained either pure oxygen gas, or a mixture of oxygen and nitrogen. The relative quantity of the latter was ascertained by the eudiometer. The knowledge of the weight of the substance, and that of the chlorate of potash; of the quantity of carbonic acid formed, and of the oxygen remaining, supplied all the data necessary for calculating the composition of the body analysed. That portion of the oxygen of the chlorate of potash which had disappeared, had of course formed water with the hydrogen of the substance.

The only objection to the apparatus of Gay-Lussac and Thénard was, that it made the accuracy of the results to depend too much on the dexterity of the operator. The analysis of substances containing nitrogen, moreover, by means of chlorate of potash, was not very exact, in consequence of the formation of nitrous acid; and it was obviously impossible to employ that salt in the analysis of liquid or volatile substances.

Berzelius endeavoured, and successfully, to render this method more convenient in the execution, and to diminish the number of calculations required. He placed the tube of combustion in the horizontal position, and collected the water formed. He also employed the chlorate of potash, mixed with a large quantity of common salt, by which means the combustion was rendered slower, and at the same time the advantage was gained of introducing the whole of the substance to be burned, into the tube, before commencing the combustion.

These forms of apparatus, which were applicable only to a very limited class of bodies, were greatly and most essentially improved by the use of oxide of copper, instead of chlorate of potash; which was first proposed by Gay-Lussac, and employed by him in the analysis of uric acid. At present, the superiority of oxide of copper is so generally admitted, that chlorate of potash is no longer employed. Besides oxide of copper, chromate of lead has of late been used in the analysis of many substances containing a large proportion of carbon.

De Saussure and Prout have both described forms of apparatus for the analysis of organic bodies, which differ from the original one of Gay-Lussac and Thénard, only in their form, and in the substitution of oxygen gas, and oxide of copper, instead of chlorate of potash.

The apparatus of Prout is so arranged that the substance to be analysed is burned, either alone, or mixed with some other body, in a known volume of oxygen gas, and the volume of the gas, after the combustion, is compared with its original bulk. This method is founded on the well known fact, that, when carbon is burned in oxygen gas, the carbonic acid gas produced, occupies exactly the same space as the oxygen consumed, and consequently does not alter its volume: as also, that when hydrogen unites with oxygen, for each volume of hydrogen, half a volume of oxygen disappears by the condensation of the water which is formed.

Consequently, if the substance to be burned consists of carbon, hydrogen, and oxygen, there are only three cases possible. Either the volume of the oxygen is unaltered; and, in this case, the substance contains oxygen and hydrogen in the proportions necessary to form water; or the volume of the oxygen is diminished, or it is inereased. In the latter cases, the body either contains more hydrogen, and consequently less oxygen than is sufficient to form water; or there is less hydrogen, and consequently more oxygen than is required for that purpose. The diminution or increase in the volume of the oxygen can be exactly measured, and, the quantity of the carbonic acid produced being ascertained, it is easy to express the composition of the substance in numbers.

But this apparatus cannot be applied to the analysis of substances containing nitrogen, nor to that of many other bodies. Lately, Brunner has constructed an apparatus on a similar principle. All these forms of apparatus, however, have been employed by their inventors alone; and as they have no advantage over the one now commonly employed, it is unnecessary to describe them minutely in this place.

GENERAL METHOD OF PROCEEDING

In the next section we shall describe the instruments and processes which are at present employed by the majority of chemists for organic analysis; and we shall here prefix some general remarks on the operations which occur in such analyses.

It will be observed, that all the parts of the apparatus used for this purpose are extremely simple, and for their employment require no especial dexterity. The essential conditions for performing a good analysis are, the greatest accuracy in weighing, and the strictest conscientiousness in the execution of all the preparatory steps of the process. Let us not flatter ourselves that we can obtain an accurate result, if any thing be neglected that can secure it. All the time and labour we bestow are thrown away, if we omit any one of the precautions which are recommended.

It is obvious that the object in view may be attained by various means, and that the methods described in the following pages are susceptible of improvement: but all the so-called improvements, hitherto proposed, only prove that their authors are ignorant of the most general principles of what a method ought to be.

Every chemist will be able, when he has acquired some experience in organic analysis, to alter the apparatus here described, in particular cases, according to his ideas, and to adapt it to the object he has in view; but it would be going too far to consider this deviation in a special case, as an improvement of the process in general, and to recommend it as such.

There is within the human mind an innate desire for improvement—hence the efforts to improve what we possess, and to discover new means of attaining the desired object. But we frequently commit the error of neglecting to test the utility of the known methods, or even to make ourselves familiar with them. We begin by deviating from the customary path; and if our efforts be crowned with success, the satisfaction we experience in the discovery we have made, leads us to overlook the circuitous nature of the route we have followed, and the difficulties we have had to overcome, which we should not have encountered on the beaten path.

In what follows, we hold to the rule of Berzelius, the most experienced chemist of our own, and probably also of all times; and of two equally good methods, we prefer the simple to the complicated one.

The first problem to be solved, in performing an organic analysis, is, to procure the substance to be analysed in the highest degree of purity. No means should be neglected to satisfy ourselves of the absence of foreign matters. The matter being supposed pure, we must attend to the difficulty of determining the weight of the body to be analysed, as one source of uncertainty in the results of the analysis, and of the variations occurring in different analyses of the same substance. All organic bodies greedily absorb moisture from the air, and thus become heavier. They must, therefore, first be deprived of all hygrometric moisture, and then weighed in such a manner that it is hardly possible for them to attract moisture during the time they are in the scale.

When we consider that an excess of water, to the amount of $^{7}/_{100}$ths or $^{8}/_{100}$ths of a grain is equivalent to a loss of twice as much carbonic acid, we cannot surely bestow too much attention on the accurate determination of the weight of the substance to be analysed.

This object may be obtained in various ways. The following apparatus gives complete security on this head. It consists of the tube A, *fig.* 12.1. The wide part below is about half an inch in diameter: the tubes *a* and *b* are barometer tubes,

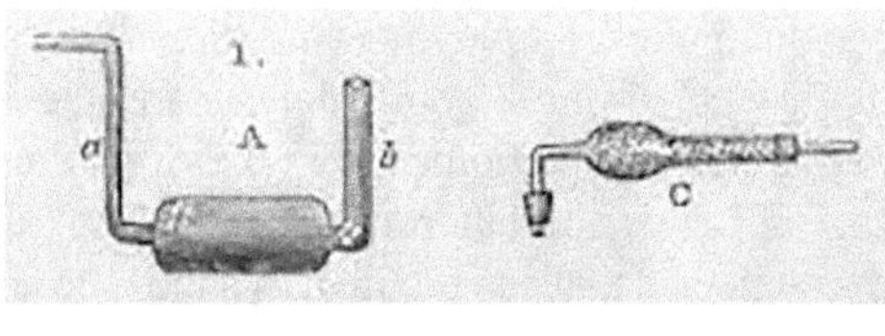

one of $^{1}/_{6}$th, the other of $^{1}/_{4}$th of an inch in diameter. The substance is introduced by the wide tube *b*, which is then connected by means of a cork with the tube C, containing fused chloride of calcium. The opposite tube *a* is joined with the tube *d, fig.* 12.2: *e* is an ordinary syphon. The tube *d* is about an inch shorter than *a'*, the external limb of the syphon.

By means of this arrangement we can produce a perfectly uniform discharge of water from the three necked bottle; and since the air, which replaces the water must enter the bottle by the tube *d*, we can instantly perceive whether all the joinings are air-tight.

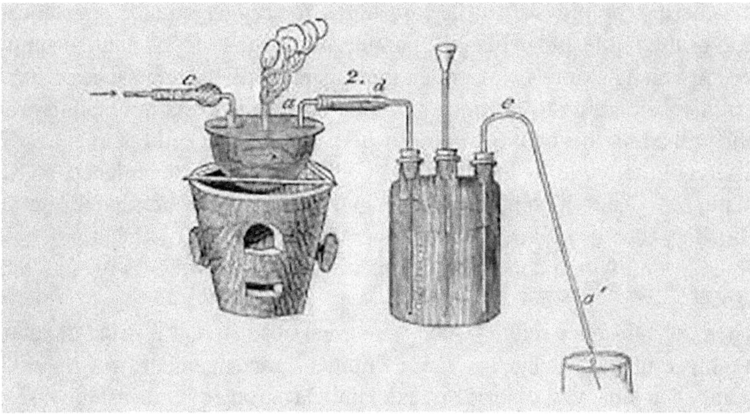

The bottle is filled with water; and, of course, when this water is made to flow out by the syphon, the continual current of dry air thus produced entirely removes all moisture from the substance.

The horizontal part of the drying apparatus is placed in a sand bath, a water bath, a bath of solution of chloride of calcium, &c., according to the temperature to which we wish to expose the substance. If we wish to determine the amount of water, the apparatus A is weighed, first empty, and then with the substance. It is then placed in the water bath, &c., and a stream of dry air is made to pass through it, as long as water condenses in the tube *d*. By weighing occasionally, we ascertain if it loses weight. When the weight becomes constant, a small portion of the substance is shaken out of *b* into a long and perfectly dry test tube, *fig.* 12.3; which is then heated by a spirit lamp or sand bath—of course to a degree insufficient to produce decomposition. If no trace of water bedew the side of the test tube, we may be certain that the substance is perfectly dry: if any moisture appear, the water bath must be replaced by one of a solution of salt, or of chloride of calcium, and the operation repeated at a higher temperature.

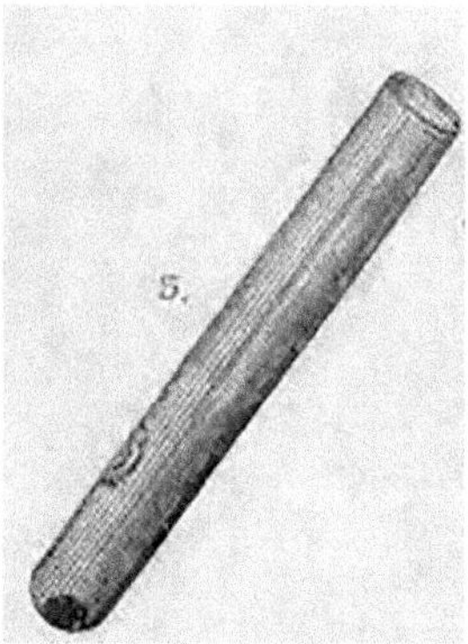

Mitscherlich employs a similar apparatus for drying organic substances; but instead of the bottle and syphon he connects with the tube *a* a hand air-pump, by means of which he draws air through the apparatus till the substance is dry. But it is extremely fatiguing to continue pumping from four to six hours; and probably no one will adopt this method, except where no three necked bottle is to be had.

Instead of the bottle we may employ, with still greater convenience, a vessel of tin plate, *fig.* 12.4, which holds about 4 gallons of water. The funnel *a* serves for replenishing the vessel when empty, when the air escapes by the middle opening *b*, which is shut with a cork at other times. The flow of water is regulated by the stopcock.

When the substance retains water with great obstinacy, it is dried in vacuo at a high temperature. The cut, *fig.* 12.5, exhibits this arrangement. A is a small hand air-pump, B a tube with chloride of calcium, D a strong cylindrical tube, sealed at

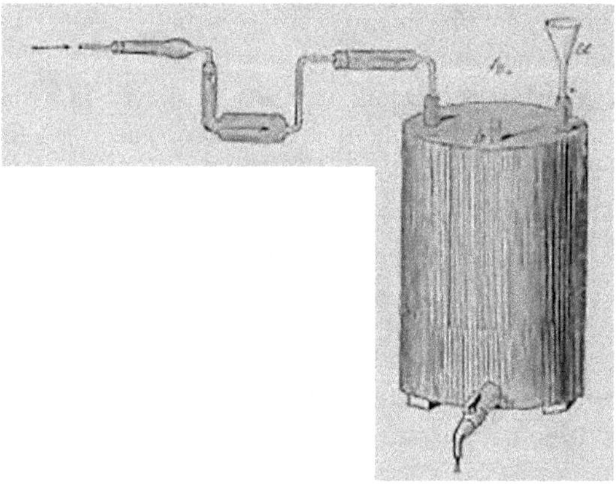

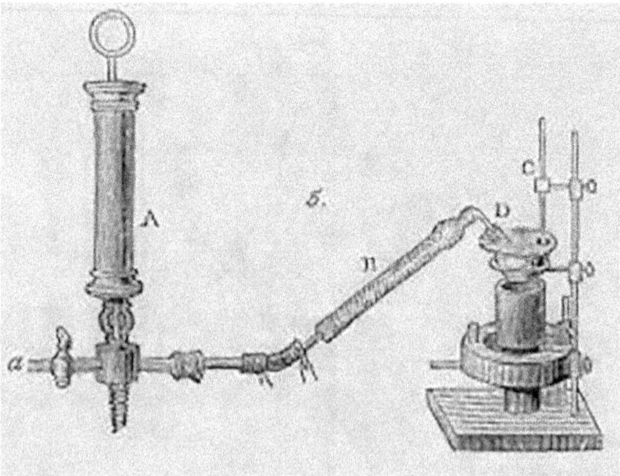

one end, which contains the substance to be dried. C is a thermometer. The tube D is placed in an iron or copper vessel with a concentrated solution of chloride of zinc, and heated nearly to the temperature at which the substance is decomposed. After the moist air has been removed by the air-pump, air is from time to time admitted into the apparatus by the stopcock *a;* this air, by passing through the chloride of calcium, is each time deprived of all hygrometric moisture, so that in a very short time, at most in a few minutes, we can by this means remove all water, whether hygrometric water or water of crystallization.

6.

When the substance is dry, a certain quantity of it must be weighed out for analysis. This is best done in a small narrow tube, open at the end. The cut, *fig.* 12.6, exhibits it of the actual size. This tube may either be placed horizontally on the scale, or set in a conical roll of tin-plate, which rests with its broad end on the pan of the balance. A stand of tin-plate is also very convenient. The tube being weighed, a portion of the substance is introduced and the tube is weighed again. The increase of weight gives the quantity of the substance.

We may also counterpoise the tube with the substance, empty the tube, and return it into the scale with any adhering particles. The weight necessary to restore the counterpoise, gives the quantity taken out of the tube.

As a general rule, we must avoid all weighing in a watch-glass, or any wide vessel. With the narrow tube recommended, no appreciable change of air can take place within it during the short time it lies on the scale; and even when containing highly hygroseopic substances, this simple apparatus does not increase in weight in the course of half an hour.

We have now a known quantity of the substance. To determine the amount of its carbon and hydrogen, we must convert the carbon into carbonic acid, and the hydrogen into water, and we must ascertain the weight of both these products.

In general, the substance, if dry and pulverizable, is mixed with oxide of copper, and the mixture heated in a glass tube by means of a charcoal fire. The tube of combustion is 15 to 18 inches long, and ⅓ to ½ of an inch in diameter. The closed end is drawn out to a point, which is bent obliquely upwards, and scaled. During the mixture of the substance with the oxide of copper, both bodies attract moisture from the air. This water would increase the weight of that formed in the combustion, and must be most carefully and completely removed before the combustion.

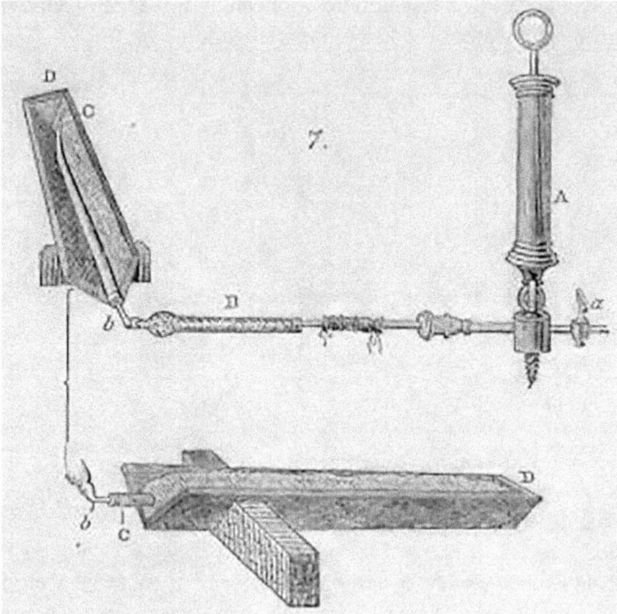

This is most easily effected by an arrangement similar to that above described, for drying the substance in vacuo at a high temperature. A, *fig.* 12.7, is the air-pump, B the tube with fused chloride of calcium, C the tube of combustion filled with the mixture, which is placed in a wedge shaped box of wood, D, and surrounded with sand heated to 250° Fahrenheit. Before pumping out the air, the tube with the mixture, C, must be subjected, in the horizontal position, to several smart taps on a table, so that a space is seen to be empty above the mixture all along the tube. If this method of furnishing a free exit to the air when pumped out be neglected, as soon as the pump is worked, a part of the mixture is forced into the tube with chloride of calcium. We now produce a vacuum in the tube, and allow dry air from time to time to enter by the stopcock, and after pumping out the air ten or twelve times, we can perceive no further deposition of moisture at the point *b* of the tube B, even when we surround it with cotton wool, and cool it by dropping ether upon it. The mixture may then be considered dry.

The substance is mixed with pure oxide of copper in a clean and warm mortar of Wedgewood's ware or porcelain. The more carefully we divide the substance and mix it with the oxide of copper, the more easy and complete is the combustion.

Mitscherlich leaves the point of the tube of combustion open, joins either end with a tube containing chloride of calcium, with the other end of which is connected a pair of bellows; heats the oxide of copper, which has been introduced so as to fill half of the tube, to a low red heat, and forces, by means of the bellows, dry air over the hot oxide. The point is then sealed up. He now counterpoises the tube with the oxide of copper, and shakes out of the tube in which it has been dried, a portion of the substance to be analysed, into the counterpoise tube. The increase of weight gives the quantity of the substance.

He now mixes it with the oxide of copper in the following manner:—He bends one end of a long copper wire into the form of a corkscrew, screws this half way into the layer of oxide, and moves it up and down till the mixture appears sufficiently intimate.

This method is less convenient and more troublesome than the one above described. In the first place the weight of the tube with oxide of copper, (from 1800 to 2100 grains,) does not permit us to determine the weight of the substance to be analysed to the $^1/_{500}$th of a grain. This is a source of uncertainty. Again, we cannot effect a complete mixture by means of a cork-screw wire. This is easily shown. If we mix some starch as intimately as possible with oxide of copper in this manner, and press the mixture in a mortar with the pestle, we easily recognise the unmixed or cohering particles of starch by a number of round white spots. The interior of these particles would only be charred, not burned. Berzelius recommends this method, when we have not an air-pump; but in that ease it would be better not to attempt the analysis of organic bodies at all.

The water formed in the combustion is collected in the tube, *fig.* 12.8, which is filled with chloride of calcium—large fragments being placed in the bulb, and coarse powder in the long tube. Near the two open ends of this tube, at *a* and *b*, is put a little cotton-wool, in order to prevent any chloride of calcium from falling out. The tube *b* is fitted tight into the larger tube by means of a cork: the cork is cut off close to the tube, and covered with sealing wax to prevent dust from adhering to it. The whole is weighed. Its increase of weight after the combustion gives the quantity of water produced.

The tube with chloride of calcium, thus prepared, is connected at the end *a* by means of a cork to the tube of combustion, *fig.* 12.9. The cork should be perforated by means of Mohr's cork-borer, and the aperture filed very smooth, and made to fit the end *a* very tight. It is then, with a very sharp straight-edged knife, cut so as to fit closely into the tube of combustion. According to the shape of the tube, the cork should be cut cylindrical or slightly conical. We must avoid

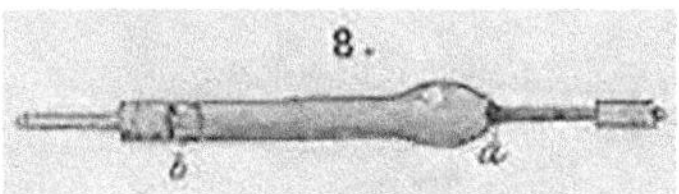

perforating the cork with a hot wire, as most corks perforated in this way crack, and become useless.

Some Chemists, according to the method of Berzelius, give to the tube with chloride of calcium, the shape represented in *fig.* 12.10. The tube of combustion *a* is drawn out into a point, which is introduced into the tube *b* with chloride of calcium. The junction is completed by a tube of caoutehoue, tied on both sides. After the combustion, the point of the tube *a* is cut through at *c*, the caoutehoue connecter removed, without taking away the point of *a* from the tube *b; b* is then weighed with the point; the point is taken out, washed, ignited, and weighed. Deducting the weight of the point from the whole increase of weight of *b*, we obtain the true increase of weight due to the water produced.

The carbonic acid formed in the combustion is collected in the apparatus, *fig.* 12.11, which is filled with a solution of caustic potash, in such a manner that a small bubble of air remains in each bulb. This apparatus consists of a glass tube, in which 5 bulbs are blown. It is made in the following manner. Two barometer tubes, *b b, fig.* 12.12, of $^1/_6$ of an inch wide, are soldered to a stout glass tube, *a*, 3 inches long and $^1/_3$ of an inch wide. One end being stopped up, *a* is softened at

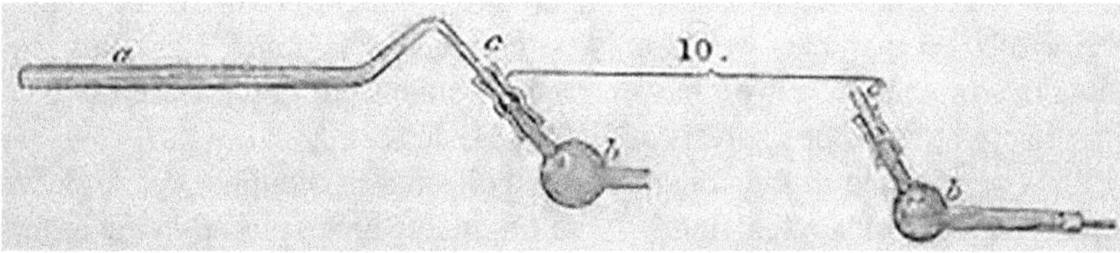

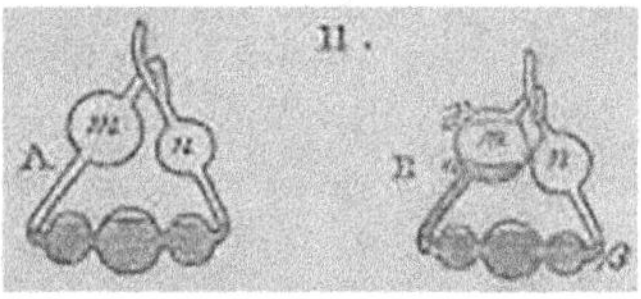

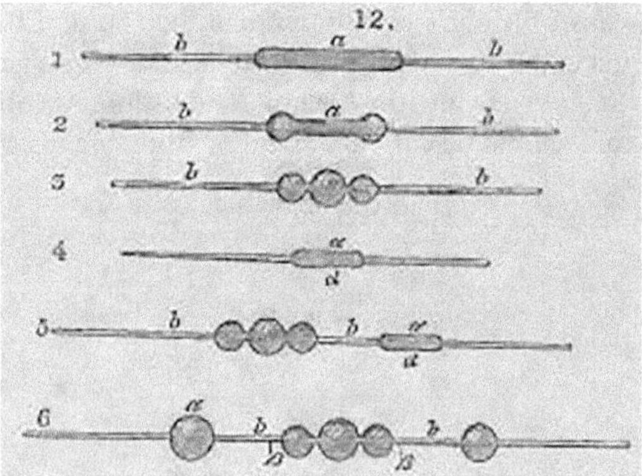

one extremity, and blown out to a bulb. The same is done at the opposite end of *a*, as may be seen in No. 2. Next, the middle part of *a* is made very hot, and blown to a somewhat larger bulb, so that the whole acquires the shape of No. 3. About an inch of a tube, *α*, equal to *a*, is now soldered to a narrow barometer tube similar to *b b;* one end of *α* is drawn out to a point, which is cut off near to *d*, and soldered to one of the points *b*—which is previously reduced to 2 inches (No. 5). *α* is now to be blown out to a bulb, (No. 6.) The same is done at the opposite end of *β β*.

One of the bulbs *α α* may be rather smaller than the other; one however must be so large as to hold more liquid than the middle bulb of *a*. The tubes are now heated by a spirit-lamp close to the bulbs, *β β*, and are bent at an angle of 45° (*fig.* 2.13.) Above the bulbs, *m* and *n*, these two tubes are again bent. Here it is to be observed, that the tube *o*, having the larger bulb *m*, should be so bent that its horizontal portion, connected with the tube containing chloride of calcium, is to the left of the operator. The tubes *o* and *p*, after being sufficiently heated at *α*, are bent so as to cross each other, as shown at B. By attending closely to the figure, we may be certain of giving to the apparatus the best form and proportions. The sharp ends of the tubes *o* and *p* must be rounded in the flame of the spirit-lamp.

[The Apparatus thus prepared has the exact form and size represented by *fig.* 12.14, page 11.—W. G.]

To fill this apparatus with caustic ley, one end must be connected by a cork with the sucker, *fig.* 12.15; the other end of the potash apparatus is then dipped into a glass containing the ley, and the liquid drawn by suction with the mouth into the apparatus.

The end of the tube which has been moistened with the ley is to be dried externally, and internally: the latter is easily accomplished by means of a small roll of filtering paper. The apparatus, when quite dry and clean, is weighed, and then joined to the tube with the chloride of calcium by means of a small tube of caoutchouc.

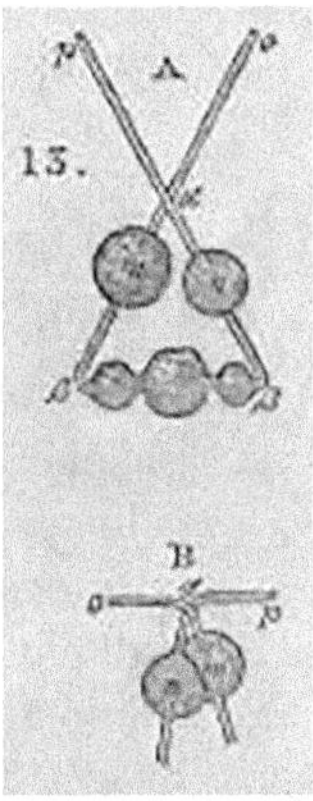

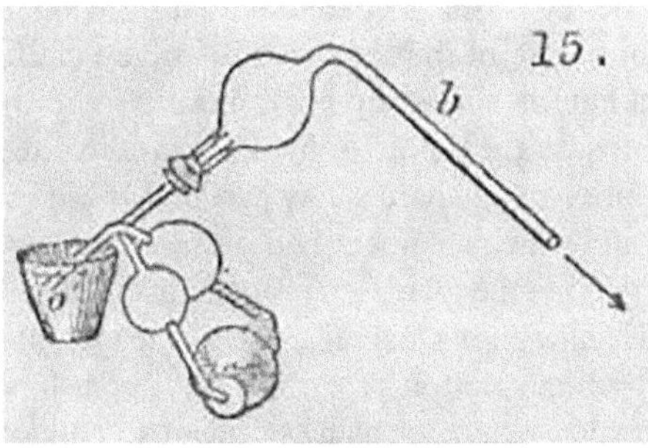

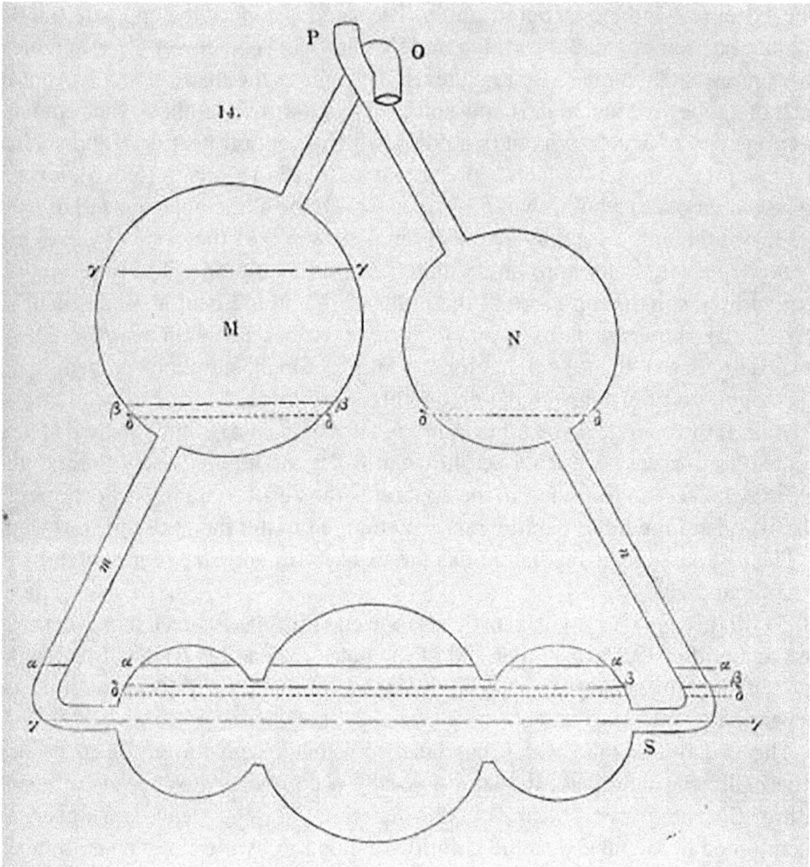

The potash apparatus, when filled with ley, commonly weighs from 750 to 900 grains. If the ley have the sp. gr. 1.25 to 1.27, it does not produce foam when bubbles of gas pass through it; and when of this strength, its absorptive power is likewise greatest. Caustic soda ley foams like soap and water, and must therefore be avoided.

[The ley is best prepared in the following manner. Two parts of the subcarbonate of potash of the shops are dissolved in 20 to 24 of boiling water. One part of quicklime is slaked by being *covered* with hot water in any convenient vessel. In this way the whole of the lime is converted into a uniform ercam, without the formation of any hard sandy particles, which oecur in the ordinary method of slaking, and which are not only useless but hurtful, by preventing, from the space they occupy, and the increased proportion of lime they render necessary, the separation, by decanting, of much of the ley. The cream of lime is added in small portions to the boiling solution of carbonate of potash, which is boiled a few minutes after each addition, water being occasionally added to supply the loss by evaporation.

When the whole of the lime has been added, the mixture is boiled for a short time longer, and is then allowed to cool in the pan or goblet, carefully closed with its lid. After 12 hours, nearly the whole of the ley may be decanted perfectly clear, and quite caustic, especially if the vessel has been nearly full. The carbonate of lime, when this process is followed exactly, is sandy, and occupies a very small bulk. The clear liquid is now to be rapidly boiled down in a clean iron vessel, till small crystals begin to separate. It is then allowed to cool in a stoppered bottle of green glass, when it deposits the whole of the sulphate of potash originally present in the subcarbonate; that salt being absolutely insoluble in a strong solution of caustic potash. For the above essential improvements in the preparation of caustic potash, we are indebted to the author of this treatise and to Dr. F. Mohr of Cobleutz. I find that the solution which has deposited the sulphate of potash possesses the sp. gr. 1.25, and is perfectly adapted for organic analysis. As all contact with organic matter has been avoided, it is also in general colourless, and yields solid caustic potash almost white, containing no impurity except a little chloride of potassium; which of course may be avoided by using genuine salt of tartar; but which does not in the least affect the use of the potash for most purposes. The necessity for using at least 10 or 12 parts of water to 1 of carbonate of potash arises from the curious fact, noticed by Professor Liebig, that when less water is present, the potash takes back the carbonic acid from the carbonate of lime.—W. G.]

The tubes of caoutchouc are made out of thin sheets of that substance. A portion, 1½ inch long, is doubled up so as to form a tube of the size of those which it is to connect. About a line in width is now to be cut off with a very clean and sharp pair of scissors, along the length of the caontchouc where the two sides meet. We thus obtain two smooth cut surfaces, which, if pressed together by the thumb nails, adhere so as to form a perfect junction. The tube is now pulled lengthways, so as to stretch it, several times. If we touch the fresh cut surfaces with the fingers, they do not cohere where they have been touched. It is right also to moisten the inside of the caoutchouc, before forming the tube, that its sides may not cohere. The caoutchouc tubes are fastened over the glass tubes by strong threads of silk, knotted at the ends, to prevent them from slipping.

[It is more effectual to dust the inside of the tube with any fine powder, such as flour or starch—removing all that is superfluous.—W. G.]

The furnace in which the combustion is carried on, is exhibited by *fig.* 12.16. It is made of sheet iron, 22 to 24 inches long, and 3 inches high. The bottom is 3 inches wide, and furnished with apertures which form a sort of grate; these apertures are narrow slits, running across, at an half inch distance from each other. The sides of

16.

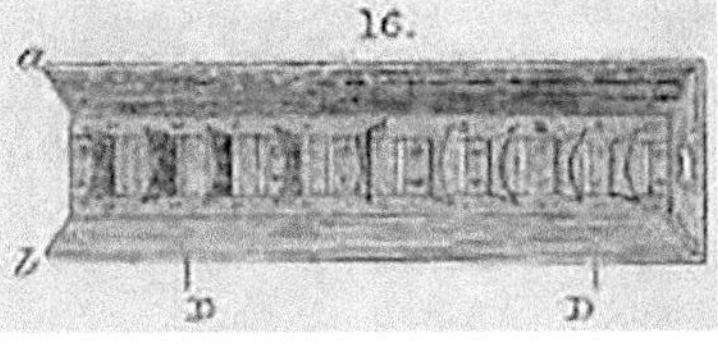

the furnace are inclined outwards so that at the top they are 4½ inches apart. The whole rests on a large tile or paving stone, *e, fig.* 12.19, in such a manner that the two slits nearest the front are left open, while the remainder are closed by the tile. Within the furnace are placed at intervals supports of strong sheet-iron, of the form D, *fig.* 12.17. They must be of equal height, and must correspond with the round aperture in the front of the furnace, A, *fig.* 12.17. Their use is to support the tube of combustion.

When we wish to increase the heat, the furnace is raised a little on one side, and a thin bit of tile is introduced at two places. This allows air to enter by all the slits in the bottom of the furnace. Good charcoal is the only fuel employed in this furnace.

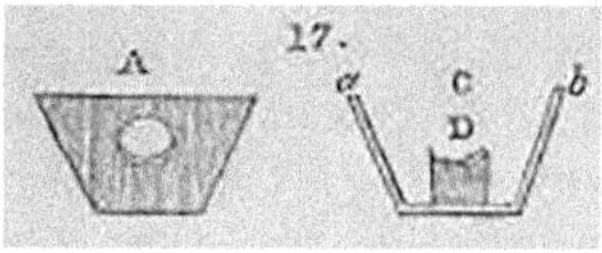

Editorial Headnote

Justus von Liebig, *Animal Chemistry, or Organic Chemistry in Its Application to Physiology and Pathology* (New York: Wiley, 1848), ix–xix

Here Liebig further highlights the importance of the relation between curiosity (our questions) and the experimental method, which 'intersect' to lead to truth.

13

ANIMAL CHEMISTRY, OR ORGANIC CHEMISTRY IN ITS APPLICATION TO PHYSIOLOGY AND PATHOLOGY

Justus von Liebig

Source: New York: Wiley, 1848, ix–xix

By the application to Chemistry of the methods which had for centuries been followed by philosophers in ascertaining the causes of natural phenomena in physics—by the observation of weight and measure—Lavoisier laid the foundation of a new science, which, having been cultivated by a host of distinguished men, has, in a singularly short period, reached a high degree of perfection.

It was the investigation and determination of all the conditions which are essential to an observation or an experiment, and the discovery of the true principles of scientific research, that protected chemists from error, and conducted them, by a way equally simple and secure, to discoveries which have shed a brilliant light on those natural phenomena which were previously the most obscure and incomprehensible.

The most useful applications to the arts, to industry, and to all branches of knowledge related to chemistry, sprung from the laws thus established; and this influence was not delayed till chemistry had attained its highest perfection, but came into action with each new observation.

All existing experience and observation in other departments of science reacted, in like manner, on the improvement and development of chemistry; so that chemistry received from metallurgy and from other industrial arts as much benefit as she had conferred on them. While they simultaneously increased in wealth, they mutually contributed to the development of each other.

After mineral chemistry had gradually attained its present state of development, the labors of chemists took a new direction. From the study of the constituent parts of vegetables and animals, new and altered views have arisen; and the present work is an attempt to apply these views to physiology and pathology.

DOI: 10.4324/9781003009337-19

In earlier times the attempt has been made, and often with great success, to apply to the objects of the medical art the views derived from an acquaintance with chemical observations. Indeed, the great physicians, who lived towards the end of the seventeenth century, were the founders of chemistry, and in those days the only philosophers acquainted with it. The phlogistic system was the dawn of a new day; it was the victory of philosophy over the rudest empiricism.

With all its discoveries, modern chemistry has performed but slender services to physiology and pathology; and we cannot be deceived as to the cause of this failure, if we reflect that it was found impossible to trace any sort of relation between the observations made in inorganic chemistry, the knowledge of the characters of the elementary bodies and of such of their compounds as could be formed in the laboratory, on the one hand, and the living body, with the characters of its constituents, on the other.

Physiology took no share in the advancement of chemistry, because for a long period she received from the latter science no assistance in her own development. This state of matters has been entirely changed within five-and-twenty years. But during this period physiology has also acquired new ways and methods of investigation within her own province; and it is only the exhaustion of these sources of discovery which has enabled us to look forward to a change in the direction of the labours of physiologists. The time for such a change is now at hand; and a perseverance in the methods lately followed in physiology would now, from the want, which must soon be felt, of fresh points of departure for researches, render physiology more extensive, but neither more profound nor more solid.

No one will venture to maintain, that the knowledge of the forms and of the phenomena of motion in organized beings is either unnecessary or unprofitable. On the contrary, this knowledge must be considered as altogether indispensable to that of the vital processes. But it embraces only one class of the conditions necessary for the acquisition of that knowledge, and is not of itself sufficient to enable us to attain it.

The study of the uses and functions of the different organs, and of their mutual connection in the animal body, was formerly the chief object of physiological researches; but lately this study has fallen into the back-ground. The greater part of all the modern discoveries has served to enrich comparative anatomy far more than physiology.

These researches have yielded the most valuable results in relation to the recognition of the dissimilar forms and conditions to be found in the healthy and in the diseased organism; but they have yielded no conclusions calculated to give us a more profound insight into the essence of the vital processes.

The most exact anatomical knowledge of the structure of the tissues cannot teach us their uses; and from the microscopical examination of the most minute reticulations of the vessels we can learn no more as to their functions than we have learned concerning vision from counting the surfaces on the eye of the fly. The most beautiful and elevated problem for the human intellect, the discovery of the laws of vitality, cannot be resolved, nay, cannot even be imagined, without an accurate knowledge of chemical forces; of those forces which do not act at sensible distances; which are manifested in the same way as those ultimate causes by which the vital phenomena are determined; and which are invariably found active, whenever dissimilar substances come into contact.

Physiology, even in the present day, still endeavors, but always after the fashion of the phlogistic chemists (that is, by the *qualitative* method,) to apply chemical experience to the removal of diseased conditions; but with all these countless experiments we are not one step nearer to the causes and the essence of disease.

Without proposing well-defined questions, experimenters have placed blood, urine, and all the constituents of the healthy or diseased frame, in contact with acids, alkalies, and all sorts of chemical re-agents; and have drawn, from observation of the changes thus produced, conclusions as to their behaviour in the body.

By pursuing this method, useful remedies or modes of treatment might by accident be discovered; but a rational physiology cannot be founded on mere reactions, and the living body cannot be viewed as a chemical laboratory.

In certain diseased conditions, in which the blood acquires a viscid consistence, this state cannot be permanently removed by a chemical action on the fluid circulating in the blood-vessels. The deposit of a sediment from the urine may, perhaps, be prevented by alkalies, while their action has not the remotest tendency to remove the cause of disease. Again, when we observe, in typhus, insoluble salts of ammonia in the fæces, and a change in the globules of the blood similar to that which may be artificially produced by ammonia, we are not, on that account, entitled to consider the presence of ammonia in the body as the cause, but only as the effect of a cause.

Thus medicine, after the fashion of the Aristotelian philosophy, has formed certain conceptions in regard to nutrition and sanguification; articles of diet have been divided into nutritious and non-nutritious; but these theories, being founded on observations destitute of the conditions most essential to the drawing of just conclusions, could not be received as expressions of the truth.

How clear are now to us the relations of the different articles of food to the objects which they serve in the body, since organic chemistry has applied to the investigation her *quantitative method* of research!

When a lean goose, weighing 4 lbs., gains, in thirty-six days, during which it has been fed with 24 lbs. of maize, 5 lbs. in weight and yields 3½ lbs. of pure fat, this fat cannot have been contained in the food, ready formed, because maize does not contain the thousandth part of its weight of fat, or of any substance resembling fat. And when a certain number of bees, the weight of which is exactly known, being fed with pure honey, devoid of wax, yield one part of wax for every twenty parts of honey consumed, without any change being perceptible in their health or in their weight, it is impossible any longer to entertain doubt as to the formation of fat from sugar in the animal body.

We must adopt the method which has thus led to the discovery of the origin of fat, in the investigation of the origin and alteration of the secretions, as well as in the study of all the other phenomena of the animal body. From the moment that we begin to look earnestly and conscientiously for the true answers to our questions, that we take the trouble, by means of weight and measure, to fix our observations, and express them in the form of equations, these answers are obtained without difficulty.

However numerous our observations may be, yet, if they only bear on one side of a question, they will never enable us to penetrate the essence of a natural phenomenon in its full significance. If we are to derive any advantage from them, they must be directed to a well-defined object; and there must be an organized connection between them.

Mechanical philosophers and chemists justly ascribe to their methods of research the greater part of the success which has attended their labors. The result of every such investigation, if it bear in any degree the stamp of perfection, may always be given in few words; but these few words are eternal truths, to the discovery of which numberless experiments and questions were essential. The researches themselves, the laborious experiments and complicated apparatus, are forgotten as soon as the truth is ascertained. They were the ladders, the shafts, the tools, which were indispensable to enable us to attain to the rich vein of ore; they were the pillars and air passages which protected the mine from water and from foul air.

Every chemical or physical investigation, however insignificant, which lays claim to attention, must in the present day possess this character. From a certain number of observations it must enable us to draw some conclusion, whether it be extended or limited.

The imperfection of the method or system of research adopted by physiologists can alone explain the fact, that for the last fifty years they have established so few new and solid truths in regard to a more profound knowledge of the functions of the most important organs, of the spleen, of the liver, and of the numerous glands of the body; and the limited acquaintance of physiologists with the methods of research employed in chemistry will continue to be the chief impediment to the progress of physiology, as well as a reproach which that science cannot escape.

Before the time of Lavoisier, Scheele, and Priestley, chemistry was not more closely related to physics than she is now to physiology. At the present day chemistry is so fused, as it were, into physics, that it would be a difficult matter to draw the line between them distinctly. The connection between chemistry and physiology is the same, and in another half-century it will be found impossible to separate them.

Our questions and our experiments intersect in numberless curved lines the straight line that leads to truth. It is the points of intersection that indicate to us the true direction; but, owing to the imperfection of the human intellect, these curve lines must be pursued. Observers in chemistry and physics have the eye ever fixed on the object which they seek to attain. One may succeed, for a space, in following the direct line; but all are prepared for circuitous paths. Never doubting of the ultimate success of their efforts, provided they exhibit constancy and perseverance, their eagerness and courage are only exalted by difficulties.

Detached observations, without connection, are points scattered over the plain, which do not allow us to choose a decided path. For centuries chemistry presented nothing but these points, and sufficient means were available to fill up the intervals between them. But permanent discoveries and real progress were only made when chemists ceased to make use of fancy to connect them.

My object in the present work has been to direct attention to the points of intersection of chemistry with physiology, and to point out those parts in which the sciences become, as it were, mixed up together. It contains a collection of problems, such as chemistry at present requires to be resolved; and a number of conclusions drawn according to the rules of that science from such observations as have been made.

These questions and problems will be resolved: and we cannot doubt that we shall have in that case a new physiology and a rational pathology. Our sounding line, indeed, is not long enough to measure the depths of the sea, but is not on that account less valuable to us: if it assist us, in the mean time, to avoid rocks and shoals, its use is sufficiently obvious. In the hands of the physiologist, organic chemistry must become an intellectual instrument, by means of which he will be enabled to trace the causes of phenomena invisible to the bodily sight; and if among the results which I have developed or indicated in this work, one alone shall admit of a useful application, I shall consider the object for which it was written as fully attained. The path which has led to it will open up other paths; and this I consider as the most important object to be gained.

Editorial Headnote

Justus von Liebig, 'Autobiographical Sketch', *The Popular Science Monthly*, 40 (1892–3), 655–666

On the formation of the scientific self, through curiosity, the attraction of work and a voracious appetite for knowledge.

14

'AUTOBIOGRAPHICAL SKETCH'

Justus von Liebig

Source: *The Popular Science Monthly,* 40 (1892–3), 655–666

JUSTUS VON LIEBIG: AN AUTOBIOGRAPHICAL SKETCH[1]

MY father, who had a color warehouse, frequently occupied himself in making some of the colors in which he dealt, and for that purpose had fitted up for himself a small laboratory to which I had access, and where I sometimes enjoyed the privilege of helping him. He made his experiments as prescribed in works upon chemistry, which were, with great liberality, lent to the inhabitants of Darmstadt from the rich Court Library.

The lively interest which I took in my father's labors naturally led me to read the books which guided him in his experiments, and such a passion for these books was gradually developed in me that I became indifferent to every other thing that ordinarily attracts children. Since I did not fail to fetch the books from the Court Library myself, I became acquainted with the librarian Hess, who occupied himself successfully with botany, and as he took a fancy to the little fellow, I got, through him, all the books I could desire for my own use. Of course, the reading of books went on without any system. I read the books just as they stood upon the shelves, whether from below upward or from right to left was all the same to me; my fourteen-year-old head was like an ostrich stomach for their contents, and among them I found side by side upon the shelves the thirty-two volumes of Macquer's Chemical Dictionary, Basil Valentine's Triumphal Car of Antimony, Stahl's Phlogistic Chemistry—thousands of essays and treatises in Göttling's and Gehlen's periodicals, the works of Kirwan, Cavendish, etc.

I am quite sure that this manner of reading was of no particular use so far as acquisition of exact knowledge is concerned, but it developed in me the faculty, which is peculiar to chemists more than to other natural philosophers, of thinking in terms of phenomena; it is not very easy to give a clear idea of phenomena to any one who can not recall in his imagination a mental picture of what he sees and hears, like the poet and artist for example. Most closely akin is the peculiar power of

DOI: 10.4324/9781003009337-20

the musician, who, while composing, thinks in tones which are as much connected by laws as the logically arranged conceptions in a conclusion or series of conclusions. There is in the chemist a form of thought by which all ideas become visible to the mind as the strains of an imagined piece of music. This form of thought is developed in Faraday in the highest degree, whence it arises that to one who is not acquainted with this method of thinking, his scientific works seem barren and dry, and merely a series of researches strung together, while his oral discourse, when he teaches or explains, is intellectual, elegant, and of wonderful clearness.

The faculty of thinking in phenomena can only be cultivated if the mind is constantly trained, and this was effected in my case by my endeavoring to perform, so far as my means would allow me, all the experiments whose description I read in the books. These means were very limited, and hence it arose that, in order to satisfy my inclination, I repeated such experiments as I was able to make a countless number of times, until I ceased to see anything new in the process, or till I knew thoroughly every aspect of the phenomenon which presented itself. The natural consequence of this was the development of a memory of the sense, that is to say of the sight, a clear perception of the resemblances or differences of things or of phenomena, which afterward stood me in good stead.

One will easily understand this if one imagines, for instance, a white or colored precipitate which is produced by mixing two liquids; it is formed either at once or after some time, it is cloudy or of a curdy or gelatinous character, sandy or crystalline, dull or bright, it deposits easily or slowly, etc.; or if it is colored it has a certain tint. Among the countless white precipitates each has something peculiar to itself; and when one has experience in this sort of appearances, whatever one sees during an investigation at once awakens the remembrance of what one has seen. The following example will make clear what I mean by sight or eye memory: During our joint research on uric acid, Wöhler one day sent me a crystalline body which he had obtained by the action of peroxide of lead upon this acid: I immediately thereupon wrote to him with great joy, and without having analyzed the body, that it was allantoin. Seven years before I had had this body in my hands; it had been sent to me by C. Gmelin for investigation, and I had published an analysis of it in Poggendorf's *Annalen;* since that time I had not seen it again. But when we had analyzed the substance obtained from uric acid there appeared a difference in the amount of carbon, the new body gave one and a half per cent carbon less, and since the nitrogen had been determined by the qualitative method a corresponding quantity (four per cent) of nitrogen more; consequently it could not possibly be allantoin. However, I trusted my eye-memory more than my analysis, and was quite sure that it was allantoin, and the thing now to be done was to find the remains of the substance previously analyzed in order to analyze it afresh. I could describe the little glass in which it was with such precision that my assistant at last succeeded in picking it out from among a couple of thousand other preparations. It looked exactly like our new body, except that examination under the lens showed that Gmelin, in the preparation of his allantoin, had purified it with animal charcoal, some of which having passed through the paper in the filtration had become mixed with the crystals.

Without the complete conviction which I had that the two bodies were identical, the allantoin produced artificially from uric acid would undoubtedly have been regarded as a new body, and would have been designated by a new name, and one of the most interesting relations of uric acid to one of the constituents of the urine of the fœtus of the cow would perhaps have remained for a long time unobserved. In this manner it came to pass that everything I saw remained intentionally or unintentionally fixed in my memory with equal photographic fidelity. At a neighboring soap-boiler's I saw the process of boiling soap, and learned what "curd soap" and "fitting" are, and how white soap is made; and I had no little pleasure when I succeeded in showing a piece of soap of my own making, perfumed with oil of turpentine. In the workshop of the tanner and dyer, the smith and brass-founder, I was at home, and ready to do any hand's turn.

In the market at Darmstadt I watched how a peripatetic dealer in odds and ends made fulminating silver for his pea-crackers. I observed the red vapors which were formed when he dissolved his silver, and that he added to it nitric acid, and then a liquid which smelt of brandy, and with which he cleaned dirty coat-collars for the people. With this bent of mind it is easy to understand that my position at school was very deplorable; I had no ear-memory and retained nothing or very little of what is learned through this sense; I found myself in the most uncomfortable position in which a boy could possibly be; languages and everything that is acquired by their means, that gains praise and honor in the school, were out of my reach; and when the venerable rector of the gymnasium (Zimmermann), on one occasion of his examination of my class, came to me and made a most cutting remonstrance with me for my want of diligence, how I was the plague of my teachers and the sorrow of my parents, and what did I think was to become of me, and when I answered him that I would be a chemist, the whole school and the good old man himself broke into an uncontrollable fit of laughter, for no one at the time had any idea that chemistry was a thing that could be studied.

Since the ordinary career of a gymnasium student was not open to me, my father took me to an apothecary at Heppenheim in the Hessian Bergstrasse; but at the end of ten months he was so tired of me that he sent me home again to my father. I wished to be a chemist, but not a druggist. The ten months sufficed to make me completely acquainted alike with the use and the manifold applications of the thousand and one different things which are found in a druggist's shop.

Left to myself in this way, without advice and direction, I completed my sixteenth year, and my persistent importunity at last induced my father to give me permission to go to the University of Bonn; whence I followed to Erlangen the Professor of Chemistry, Kastner, who had been called to the Bavarian University. There arose at that time at the newly established University of Bonn an extraordinary quickening of scientific life; but the degenerate philosophical methods of investigation, as they had been embodied in Oken, and still worse in Wilbrand, had a most pernicious influence on the branches of natural science, for it had led alike in lecture and in study to a want of appreciation of experiment and of an unprejudiced observation of Nature, which was ruinous to many talented young men.

From the professional chair the pupil received an abundance of ingenious contemplations; but, bodiless as they were, nothing could be made of them. The lectures of Kastner, who was considered a most eminent chemist, were without order, illogical, and arranged just like the jumble of knowledge which I carried about in my head. The relations which he discovered between phenomena were somewhat after the following pattern:

"The influence of the moon upon the rain is clear, for as soon as the moon is visible the thunderstorm ceases," or "the influence of the sun's rays on water is shown by the rise of the water in the shafts of mines, some of which can not be worked in the height of summer." That we see the moon when the thunderstorm is dispelled, and that the water rises in the mine when the brooks which drive the pumps dry up in summer, was, of course, too blunt an explanation for a clever lecture.

It was then a very wretched time for chemistry in Germany. At most of the universities there was no special chair for chemistry; it was generally handed over to the professor of medicine, who taught it, as much as he knew of it, and that was little enough, along with the branches of toxicology, pharmacology, materia medica, practical medicine, and pharmacy. Many years after this in Giessen, descriptive and comparative anatomy, physiology, zoölogy, natural history, and botany were in one single hand.

While the labors of the great Swedish chemist, the English and French natural philosophers, Humphry Davy, Wollaston, Biot, Arago, Fresnel, Thenard, and Dulong, opened up entirely new spheres of investigation, all these inestimable acquisitions found no soil in Germany where they could bear fruit. Long years of war had undermined the well-being of the people, and external political pressure had brought in its train the desolation of our universities, filled men with painful anxiety for many years, and turned their desires and their strength in other directions. The national spirit had asserted its freedom and independence in ideal spheres, and by the destruction of belief in authority had brought rich blessings in many ways—for example, in medicine and philosophy; only in physiology it had broken through its natural limits, and wandered far beyond experience.

The goal of science and the fact that it has value only when it is useful to life had almost dropped out of sight, and men amused themselves in an ideal world which had no connection with the real one. It was considered an almost debasing sentiment, and one unworthy of an educated person, to believe that in the body of a living being the crude and vulgar inorganic forces played any part. Life and all its manifestations and conditions were perfectly clear. Natural phenomena were clothed in bewitchingly lovely dress, cut out and fitted by clever men, and this was called philosophical investigation. Experimental instruction in chemistry was all but extinct at the universities, and only the highly educated pharmacists, Klaproth, Hermbstädt, Valentin Rose, Trommsdorff, and Buchholz, had themselves preserved it, but in another department.

I remember, at a much later period, Prof. Wurzer, who held the chair of chemistry at Marburg, showing me a wooden table drawer, which had the property of producing quicksilver every three months. He possessed an apparatus which mainly consisted of a long clay pipe-stem, with which he converted oxygen into nitrogen by making the porous pipe-stem red hot in charcoal, and passing oxygen through it.

Chemical laboratories, in which instruction in chemical analysis was imparted, existed nowhere at that time. What passed by that name were more like kitchens filled with all sorts of furnaces and utensils for the carrying out of metallurgical or pharmaceutical processes. No one really understood how to teach it. I afterward followed Kastner to Erlangen, where he had promised to analyze some minerals with me; but unfortunately he did not himself know how to do it, and he never carried out a single analysis with me.

The benefit which I gained through intercourse with other students during my sojourn in Bonn and Erlangen was the discovery of my ignorance in very many subjects which they brought with them from school to the university, and since I got nothing to do in chemistry I laid out all my energies to make up for my previously neglected school studies. In Bonn and Erlangen small numbers of students joined with me in a chemico-physical union, in which every member in turn had to read a paper on the question of the day, which, of course, consisted merely in a report on the subjects of the essays which appeared monthly in Gilbert and Schweigger's *Journal*.

In Erlangen, Schelling's lectures attracted me for a time, but Schelling possessed no thorough knowledge in the province of natural science, and the dressing up of natural phenomena with analogies and in images, which was called exposition, did not suit me. I returned to Darmstadt fully persuaded that I could not attain my ends in Germany.

The dissertations of Berzelius—that is to say, the better translation of his handbook, which had a large circulation at that time—were as springs in the desert. Mitscherlich, H. Rose, Wöhler, and Magnus had then repaired to Berzelius, in Stockholm; but Paris offered me means of instruction in many other branches of natural science, as, for instance, physics, such as could be found united in no other place. I made up my mind to go to Paris. I was then seventeen and a half years old. My journey to Paris, the way and manner in which I came in contact with Thenard, Humboldt, Dulong, and with Gay-Lussac, and how the boy found favor in the sight of those men, borders on the fabulous, and would be out of place here. Since then it has frequently been my experience that marked talent awakens in all men, I believe I may say without exception, an irrepressible desire to bring about its development. Each helps in his own way, and all together as if they were acting in concert; but talent only compels success if it is united with a firm, indomitable will. External hindrances to its development are in most cases very much less than those which lie in men themselves; for just as no one of the forces of Nature, however mighty it may be, ever produces an effect by itself alone, but always only in conjunction with other forces; so a man can only make valuable that which he learns without trouble, or acquires readily, for which, as we say, he has a natural gift, if he learns many other things in addition, which perhaps cost him more trouble to acquire than other people.

Lessing says that talent really is *will and work*, and I am very much inclined to agree with him.

The lectures of Gay-Lussac, Thenard, Dulong, etc., in the Sorbonne, had for me an indescribable charm; the introduction of astronomical or mathematical method into chemistry, which changes every problem when possible into an equation, and assumes in every uniform sequence of two phenomena a quite certain connection of cause and effect, which, after it has been searched for and discovered, is called "explanation" or "theory," had led the French chemists and physicists to their great discoveries. This kind of "theory" or "explanation" was as good as unknown in Germany, for by these expressions was understood not something "experienced," but always something which man must add on and which he fabricates.

French exposition has, through the genius of the language, a logical clearness in the treatment of scientific subjects very difficult of attainment in other languages, whereby Thenard and Gay-Lussac acquired a mastery in experimental demonstration. The lecture consisted of a judiciously arranged succession of phenomena—that is to say, of experiments whose connection was completed by oral explanations. The experiments were a real delight to me, for they spoke to me in a language I understood, and they united with the lecture in giving definite connection to the mass of shapeless facts which lay mixed up in my head without order or arrangement. The antiphlogistic or French chemistry had, it is true, brought the history of chemistry before Lavoisier to the guillotine; but one observed that the knife only fell on the shadow, and I was much more familiar with the phlogistic writings of Cavendish, Watt, Priestley, Kirwan, Black, Scheele, and Bergmann, than with the antiphlogistic; and what was represented in the Paris lectures as new and original facts appeared to me to be in the closest relation to previous facts, so much so, indeed, that when the latter were imagined away the others could not be.

I recognized, or more correctly perhaps the consciousness dawned upon me, that a connection in accordance with fixed laws exists not only between two or three, but between all chemical phenomena in the mineral, vegetable, and animal kingdoms; that no one stands alone, but each being always linked with another, and this again with another, and so on, all are connected with each other, and that the genesis and disappearance of things is an undulatory motion in an orbit.

What impressed me most in the French lectures was their intrinsic truth, and the careful avoidance of all pretense in the explanations; it was the most complete contrast to the German lectures, in which the whole scientific teaching had lost its solid construction through the preponderance of the deductive method.

An accidental occurrence drew A. von Humboldt's attention to me in Paris, and the interest which he took in me induced Gay-Lussac to complete, in conjunction with me, a piece of work which I had begun. In this manner I had the good fortune to enjoy the closest intercourse with the great natural philosopher; he worked with me as he had formerly worked with Thenard; and I can well say that the foundation of all my later work and of my whole course was laid in his laboratory in the arsenal.

I returned to Germany, where through the school of Berzelius, H. Rose, Mitscherlich, Magnus, and Wöhler, a great revolution in inorganic chemistry had already commenced. Through the support of von Humboldt's warm recommendation, an extraordinary professorship of chemistry at Giessen was conferred upon me in my twenty-first year.

My career in Giessen commenced in May, 1824. I always recall with pleasure the twenty-eight years which I spent there: it was as if Providence had led me to the little university. At a larger university or in a larger place my energies would have been divided and dissipated, and it would have been much more difficult, and perhaps impossible, to reach the goal at which I aimed; but at Giessen everything was concentrated in work, and in this I took passionate pleasure. The need for an institution in which the students could be instructed in the art of chemistry, by which I mean familiarity with chemical analytical operations, and skill in the use of apparatus, was then being felt; and hence it happened that, on the opening of my laboratory for teaching analytical chemistry and the methods of chemical research, students by degrees streamed to it from all sides. As the numbers increased I had the greatest difficulty with the practical teaching itself. In order to teach a large number at one time it was necessary to have a systematic plan, or step-by-step method, which had first to be thought out and put to the proof. The manuals which several of my pupils have published later (Fresenius and Will) contain essentially, with little deviation, the course which was followed at Giessen; it is now familiar in almost every laboratory.

The production of chemical preparations was an object to which I paid very particular attention; it is very much more important than is usually believed, and one can more frequently find men who can make very good analyses than such as are in a position to produce a pure preparation in the most judicious way. The formation of a preparation is an art, and at the same time a qualitative analysis, and there is no other way of making one's self acquainted with the various chemical properties of a body than by first producing it out of the raw material, and then converting it into its numerous compounds and so becoming acquainted with them.

By ordinary analysis one does not learn by experience what an important means of separation crystallization is in skillful hands; and just as little the value of an acquaintance with the peculiarities of different solvents. Consider only an extract of a plant or of flesh which contains half a dozen crystalline bodies in very small quantities imbedded in extraneous matter, which almost entirely masks the properties of the others; and yet, in this magma, we can recognize by means of chemical reactions the peculiarities of every single body in the mixed mass, and learn to distinguish what is a product of decomposition and what is not, in order to be able to separate them afterward by means which will exert no decomposing influence. An example of the great difficulty of finding the right way in such researches is afforded by the analysis of bile by Berzelius. Of all the numerous substances which he has described as its constituents no one is, properly speaking, contained in the natural bile.

An extremely short time had been sufficient for the famous pupils of the Swedish master to give a wonderful degree of perfection to mineral analysis, which depends on an accurate knowledge of the properties of inorganic bodies; their compounds and their behavior to each other were studied in all directions by the Swedish school with a keenness quite unusual previously and even now unsurpassed.

Physical chemistry, which investigates the uniform relations between physical properties and chemical composition, had already gained a firm foundation by the discoveries of Gay-Lussac and von Humboldt on the combining proportions of bodies in the gaseous state, and those of Mitscherlich on the relations between crystalline form and chemical composition; and in chemical proportions the structure appeared to have received its coping-stones and to stand forth completed. All that foreign countries had acquired in by-gone times in the way of discoveries now yielded rich fruit also in Germany.

Organic chemistry—or what is now called organic chemistry—had then no existence. It is true that Thenard and Gay-Lussac, Berzelius, Prout, and Döbereiner, had already laid the foundations of organic analysis, but even the great investigations of Chevreul upon the fatty bodies excited but little attention for many years. Inorganic chemistry demanded too much attention, and, in fact, monopolized the best energies.

The bent which I acquired in Paris was in a quite different direction. Through the work which Gay-Lussac had done with me upon fulminating silver I was familiar with organic analysis, and I very soon saw that all progress in organic chemistry depended essentially upon its simplification; for in this branch of chemistry one has to do not with different elements which can be recognized by their peculiar properties, but always with the same elements whose relative proportions and arrangement determine the properties of organic compounds. In organic chemistry an analysis is necessary to do that for which a reaction suffices in inorganic chemistry. The first years of my career in Giessen I devoted almost exclusively to the improvement of the methods of organic analysis, and the immediate result was that there began at this little university an activity which had never before been seen.

For the solution of innumerable questions connected with plants and animals, on their constituents, and on the reactions accompanying their transformation in the organism, a kindly fate brought together the most talented young men from all the countries of Europe, and any one can imagine what an abundance of facts and experiences I gained from so many thousands of experiments and analyses, which were carried out every year, and for so many years, by twenty and more indefatigable and skilled young chemists.

Actual teaching in the laboratory, of which practiced assistants took charge, was only for the beginners; the progress of my special students depended on themselves. I gave the task and supervised the carrying out of it; as the radii of a circle have all their common center. There was no actual instruction; I received from each individual every morning a report upon what he had done on the previous day, as well as his views on what he was engaged upon. I approved or made my criticisms. Every one was obliged to follow his own course. In the association and constant intercourse with each other, and by each participating in the work of all, every one learned from the others. Twice a week, in winter, I gave a sort of review of the most important questions of the day; it was mainly a report on my own and their work combined with the researches of other chemists.

We worked from break of day till nightfall. Dissipations and amusements were not to be had at Giessen. The only complaint, which was continually repeated, was that of the attendant (Aubel), who could not get the workers out of the laboratory in the evening, when he wanted to clean it. The remembrance of this sojourn at Giessen awakened in most of my pupils, as I have frequently heard, an agreeable sense of satisfaction for well-spent time.

I had the great good fortune, from the commencement of my career at Giessen, to gain a friend of similar tastes and similar aims, with whom, after so many years, I am still knit in the bonds of warmest affection. While in me the predominating inclination was to seek out the points of resemblance in the behavior of bodies or their compounds, he possessed an unparalleled faculty of perceiving their differences. A keenness of observation was combined in him with an artistic dexterity, and an ingeniousness in discovering new means and methods of research or analysis such as few men possess. The achievement of our joint work upon uric acid and oil of bitter almonds has frequently been praised; it was his work. I can not sufficiently highly estimate the advantage which the association with Wöhler brought to me in the attainment of my own as well as our mutual aims, for by that association were united the peculiarities of two schools—the good that was in each became effective by co-operation. Without envy and without jealousy, hand in hand, we pursued our way; when the one needed help the other was ready. Some idea of this relationship will be obtained if I mention that many of our smaller pieces of work which bear our joint names were done by one alone; they were charming little gifts which one presented to the other.

After sixteen years of the most laborious activity I collected the results gained, so far as they related to plants and animals, in my Chemistry applied to Agriculture and Physiology, two years later in my Animal Chemistry, and the researches made in other directions in my Chemical Letters. The last-mentioned was generally received as a popular work, which, to those who study it more closely, it really is not, or was not at the time when it appeared. Mistakes were made, not in the facts, but in the deductions about organic reactions; we were the first pioneers in unknown regions, and the difficulties in the way of keeping on the right path were sometimes insuperable. Now, when the paths of research are beaten roads, it is a much easier matter; but all the wonderful discoveries which recent times have brought forth were then our own dreams, whose realization we surely and without doubt anticipated.

Here the manuscript ends, and it is to be hoped that more of it will yet be found.

Liebig's reference to Wöhler is very touching, and shows a side of his character which all his pupils knew well; they tell many genial stories illustrating his unselfishness and kindness of heart. One could have wished that he had not considered the stories "bordering on the fabulous," of how he "found favor in the sight of Humboldt, Gay-Lussac, and Thenard, out of place here." They would have been far from out of place. Mr. Muspratt supplies one of these stories as he heard it from Liebig's own lips, in the Munich Laboratory, as follows:

Liebig frequently spoke in most grateful terms of the kind manner in which he—a youth barely eighteen—was received by Gay-Lussac, Thenard, and other eminent chemists, in Paris.

In the summer of 1823 he gave an account of his analysis of fulminating silver before the Academy. Having finished his paper, as he was packing up his preparations, a gentleman came up to him and questioned him as to his studies and future plans, and, after a most exacting examination, ended by asking him to dinner on the following Sunday. Liebig accepted the invitation, but, through nervousness and confusion, forgot to ask the name and address of his interviewer. Sunday came, and poor Liebig was in despair at not being able to keep his engagement.

The next day a friend came to him and said, "What on earth did you mean by not coming to dine with von Humboldt yesterday, who had invited Gay-Lussac and other chemists to meet you?" "I was thunderstruck," said Liebig, "and rushed off, as fast as I could run, to von Humboldt's lodgings, and made the best excuses I could." The great traveler, satisfied with the explanation, told him it was unfortunate, as he had several members of the Academy at his house to meet him, but thought he could make it all right if he would come to dinner next Sunday. He went, and there made the acquaintance of Gay-Lussac, who was so struck with the genius and enthusiasm of the youth that he took him into his private laboratory, and continued, in conjunction with him, the investigation of the fulminating compounds.—*Chemical News*.

Note

1 Read at a joint meeting of societies in the Chemical Laboratories, University College, Liverpool, on Wednesday evening, March 18, 1891, by Prof. J. Campbell Brown, D. Sc.

[At the recent celebration of the Jubilee of the Chemical Society, reference was made to the wonderful energy and ability of Liebig, to the great work which he did in founding organic chemistry, and to the immense stimulus which he gave, alike in his own country and in England, to scientific investigation in pure chemistry and in its applications to agriculture, physiology, and pathology.

Very opportunely a portion of an autobiographical sketch in Liebig's own handwriting has just come to light, in which he gives a most interesting account of the formation of his habits of thought, and of the development of his scientific activity. He also gives an amusing description of the lectures given in his student days by professors of the deductive method.

In his sixtieth year, we are told, Liebig wrote some biographical sketches which were laid aside and could not be found when he wished to resume them. They were never finished. A portion of the manuscript was found among some other papers in Liebig's handwriting by his son Dr. Georg Baron von Liebig, and has been published by the latter in the *Deutsche Rundschau* for January, 1891. Mr. E. K. Muspratt has been good enough to lend me a copy which he received from his friend the present baron.

I have endeavored to render it into English as literally as the difference in the idiom and modes of expression in the two languages will permit; and it is now made public in England by the kind permission of the *Deutsche Rundschau*.

His method of teaching and its remarkable success are worthy of attention at the present time, when technical education is occupying so much of the public mind.]

Editorial Headnote

George Gore, *The Art of Scientific Discovery; or, The General Conditions and Methods of Research in Physics and Chemistry* (London: Longmans, Green, and Co., 1878), 306–324, 372–397

George Gore, 1826–1908, generally famed for inventing the safety match, was an electrochemist. Gore was first inspired by the medical possibilities of galvanism, later expanding his expertise in physics and chemistry. He sounds a similar note to Liebig here, promoting the link between the imagination of hypotheses and the design of experiments in order to make 'discoveries', laying out the importance of practical and intellectual dexterity.

15

THE ART OF SCIENTIFIC DISCOVERY; OR, THE GENERAL CONDITIONS AND METHODS OF RESEARCH IN PHYSICS AND CHEMISTRY

George Gore

Source: London: Longmans, Green, and Co., 1878, 306–324, 372–397

A DISCOVERER is a tester of scientific ideas; he must not only be able to imagine likely hypotheses, and to select suitable ones for investigation, but, as hypotheses may be true or untrue, he must also be competent to invent appropriate experiments for testing them, and to devise the requisite apparatus and arrangements.

Experiments and observations are the original sources of new scientific knowledge. The chief use of experiments is to enable us to test hypotheses and verify true ones; to expel error, increase the certainty of our knowledge, and extend the domain of scientific truth. As no man can foretell with certainty, except in special cases, the result of a new and untried experiment, trials and observations must be made, in order to settle the question, and wrong paths have often to be traversed before we find the right one. From every experiment, however, if carefully performed, some definite knowledge may be obtained, provided its conditions are definite and known; and by means of a sufficient variety and number of experiments we are enabled to answer a particular question, or several questions, and thus to sweep away a large number of alternate false hypotheses we had framed with it. Leonardo da Vinci, prior to Lord Bacon, suggested that the proper mode of discovering new scientific knowledge was by means of experiment and observation.

The most accurate conditions for observation are generally obtained by means of experiments, because we can usually make suitable arrangements for that purpose. Experiment is usually also a potent and immediate means of testing a question. We might, in nearly all cases, wait for ever and not meet with the conditions or results obtainable by a single good test or arrangement; for instance, the alkali metals, potassium, sodium, &c., would probably not have been seen to the present

DOI: 10.4324/9781003009337-21

day, had not special conditions been invented and arranged to set them free, and to preserve them in that state. The progress of discovery is generally more slow in subjects, in which we cannot make many experiments, such as astronomy, because we have frequently to wait a long time for the necessary conditions or phenomena to occur, as in an eclipse of Venus; and when they do happen, we are generally unable to obtain the most essential condition of a test experiment, viz. changing only one condition at a time, and are obliged to prepare everything beforehand, and make the most of the brief opportunity. Phenomena which take place in a living being during the course of a disease are also similar in those respects. Discovery is especially slow in those subjects where the phenomena not only occur only at long intervals of time, and are very brief, but are also in the highest degree uncertain as to time and place, as in the study of the phenomena of falling stones, &c.; and the same remarks apply, though to a less extent, to those of earthquakes.

Every new question which has to be tested generally requires a number of new experiments, or new modifications of old ones, and thus every experiment has to a greater or less extent to be invented. During the course of each experiment also new difficulties continually arise, and new contrivances have at once to be devised in order to overcome them, and thus ingenuity is constantly on the rack. It is therefore not only desirable that the investigator should be acquainted with all the various contrivances which have been employed by earlier investigators for overcoming their difficulties, but also necessary that he should be able to invent new ones readily himself. Such facility of invention is greatly aided by a general knowledge of all the sciences, because the ideas of the contrivances by which the questions are tested and difficulties overcome are often derived from some other branch of science different from that of the subject of research; for instance, the method of detecting the state of vibration of sounding-plates by strewing sand upon them was suggested to Chladni by Lichtenberg's experiments of scattering powders on an electrified plate of resin. Investigators, under such circumstances, often seek the aid of clever mechanics and opticians, who have often in this way, without fee or reward, rendered great and unostentatious assistance to the progress of new knowledge; a notable instance of this was the late Carl Becker, of London. In these ways also have arisen, and been gradually brought to their present state of perfection, nearly all the experimental apparatus and appliances for producing, detecting, regulating, and measuring all the different substances and forces of nature.

If the object be to ascertain by experiment a qualitative fact, and also its quantitative amount, the former is usually determined first. For instance, in chemical analysis we first make qualitative tests to ascertain what substances are present, and then proceed to find their amounts. In testing a scientific question, either the most conclusive or the most easily-made tests are generally applied the first. Having decided to test a question, and invented a method or experiment for the purpose, it is in many cases a wise course to plan and then arrange or construct a practically workable experiment or apparatus of such materials as are most ready at hand. The unessential parts of such a model may generally be of the roughest kind, but the essential conditions, as far as we know or can guess them, should be fulfilled as perfectly as possible. For example, if a feeble rotatory effect is sought for, the friction of the moving parts should be reduced to a minimum, and the motive power exalted to the highest degree. Circumstances which, we feel assured, do not affect the results, such as crudeness of some parts of the apparatus, and in some cases even impurity of the materials, may be entirely disregarded; but this point can only be determined by actual knowledge, and should not be acted upon unless we are quite certain. Eminent investigators have usually constructed the apparatus used in their first experiments on a subject in a rough sort of way; the model of the first safety-lamp made by Sir Humphry Davy was formed by wrapping a piece of wire-gauze round his thumb, and is now in the possession of the Royal Society. Undue fondness for beautiful apparatus is, in the adult student, rather a sign of an amateur than of an original mind. When an investigation is completed, an apparatus, perfect in appearance and durability as well as in action, may then with propriety be constructed.

As new phenomena, when first discovered, are often very minute, the proportions of the different parts and conditions of the experiments should be very carefully considered at the outset, or as soon as possible; for want of this, early experiments often fail, which, if properly arranged, would have succeeded. Where the effects are extremely minute, they are magnified or multiplied by means of various contrivances, which differ in almost every different case, and these contrivances are described in books relating to the various sciences. The pendulum is a well-known means of magnifying, by multiplication, minute differences of time; the mirror, for magnifying the signs of minute movements, as in the needles of magnetometers and galvanometers; the electric condenser, for multiplying the effect of minute electric charges; and so on.

Experiments are usually arranged upon a suitable degree of magnitude, such that the desired effect may be conspicuous, and interfering circumstances as small as possible, for the latter are often greater than the expected result. An ordinary chemical analysis, for example, is usually made upon from 20 to 100 grains of the substance, in order to keep within a moderate compass the unavoidable errors of the process. When the expected effect is small, the substance or apparatus for producing it should be large, and the means of detecting it should be very delicate.

Experiments are, in some cases, made for the purpose of discovering a new phenomenon, and in others for investigating an already known truth. In the latter instance, if the phenomenon appears to be of an altogether novel kind, the experiments have at the outset to be made in a less systematic way, until some guiding idea of its probable nature is obtained; but, even in such a case, the employment of a table or series of leading ideas for raising hypotheses is of advantage.[1] In a series of experiments, especially in those made for the purpose of determining causes, we usually vary only one circumstance or condition at a time, and draw from it a single or small number of conclusions. In other cases, however, by means of a single experiment we are sometimes enabled to make a whole series of determinations at once.

All experiments have limits of time; some, however, require very long periods. Sir William Thomson has in progress some experiments of diffusion of an aqueous solution of sulphate of copper in very tall vertical glass tubes, which are calculated to require several hundred years to complete them. A multitude of experiments in the subjects of crystallisation and liquid diffusion might easily be devised and commenced, which would require thousands of years for completion.

Before commencing an experiment, a plan should be formed of the mode of conducting it, *i.e.* of the order in which the various changes should be made, and of the notes to be taken. A memorandum should also be made of the different substances or apparatus employed, their conditions, forms, weights, sizes, positions, relations to each other, temperature, or other circumstances which may appear essential, in order that their effects may be studied, and that the experiment may be exactly repeated, if necessary, at any future time. In complex experiments, or in those with dangerous substances, a rehearsal is often-times necessary before the actual experiment, in order that every attention may be paid to the more critical points in the actual trial. I frequently adopted this plan whilst investigating the properties of the extremely dangerous substance, anhydrous hydrofluoric acid. Having made all the necessary preparations, a preliminary trial is then made; and, having thus found that the apparatus and materials will act, one of the first points to be determined is whether or not any of the obvious or well-known causes of interference are influencing it. If, for example, a galvanometer is used, and its needle moves, that instrument should be disconnected, and the experiment repeated without it. If the needle now moves, an interference exists, and the galvanometer must be removed to such a distance (sometimes as much as 30 or 40 feet) from the experimental apparatus that an experiment with the latter no longer affects it, except when the two are connected together. In nearly all chemical experiments, in order to prevent interferences, the first preliminary condition to be secured is a high degree of purity of the substances. Even when a new or uninvestigated phenomenon does really exist, the exclusion of interferences takes a long time, and then another long period is consumed in developing the phenomenon to a degree of magnitude suitable for investigation; and very few of the experiments made up to this point can be published, because they are imperfect.

In actual research, the experiment or apparatus is often arranged in such a way that the order and disposition of its different parts can be readily altered; and whilst the experiments are in progress, the direction, distance, and strength of the forces employed are varied, instead of preparing afresh for every modified trial. By this plan a number of results and the means of drawing many conclusions are obtained with the trouble of only one preparation. For example, if a large voltaic battery is employed, a number of experiments are prepared before charging it, and the battery is so arranged that the magnitude and tension of its current can be easily varied.

The number and extent of the preparations necessary for experiments vary, of course, with the character and magnitude of the research. Nearly always, the amount of trouble required to prepare for experiments is very much greater than that expended upon the actual trials. Waiting for substances and apparatus is also a frequent source of delay, and clearing away the residues of experiments consumes a great deal of time.

For the purposes of experiment, every physical investigator requires a suitable and sufficient supply of materials and apparatus, especially those necessary for generating, governing, detecting, and measuring the different forces of nature. Every chemical experimentalist also needs to possess a stock of ordinary and rare chemicals of the highest attainable degree of purity, as well as cruder ones for commoner purposes. Rare minerals and the residues of peculiar manufacturing processes, especially those obtained in the working of rare substances, constitute a valuable addition to the stock of an investigating chemist, because new elementary bodies and the wide-spread existence of the rarer elements have not unfrequently been discovered by examining such substances.

CHAPTER XXXIII

Necessity of manipulative skill

NEXT in importance to the skilful use of a gifted mind in research comes dexterous employment of the human hand. To the mental qualifications of scientific knowledge, imagination, and invention, it is almost indispensable to add aptitude in mechanical matters, and cleverness in experimental manipulation. Great manipulative ability can be acquired only by long practice, which should be commenced at an early age. Fondness for mechanical pursuits in a child has often betokened skill in discovery. Newton, when a youth, was clever in constructing mechanical toys. In scientific study also, as in other abstruse meditations, the mind soon becomes exhausted by intense thinking, but is usually relieved by preparing and making experiments. Study and manipulation in physics and chemistry go hand-in-hand, and in actual research they often alternate; from the results of an experiment just made we draw new conclusions and infer additional hypotheses, and we then make other experiments to test them, and so on.

'The accuracy of a determination often depends much more upon the skill of the operator than upon the construction of the instrument used; and thus Cavendish, with nitric oxide as his reagent and water as the confining liquid, made many hundred analyses of air, collected in various localities, in 1781, and found the percentage of oxygen to be invariably 20·83, a number nearly identical with those obtained by Bunsen and Regnault, with much more perfect means. But the average chemist of that day obtained the most discordant results with the same apparatus and materials, and would doubtless also do so at the present day. By improved apparatus and methods, the work of the average chemist is made to equal, or nearly so, that of the most skilled.'[1]

As every different substance possesses different properties, and every different apparatus produces different effects, or is for a more or less different purpose; and as the number of known substances is extremely great, and the kinds of apparatus extremely varied, it is evident that the modes of manipulation are equally diverse.

Some of the substances employed are very explosive, others are deadly poisonous; some are in the highest degree volatile, and require enormous pressures to liquefy them; some take fire on contact with air, others by contact with water; some explode in the presence of light, others by a slight rise of temperature; some require to be made liquid by the very highest temperature, others by the application of the greatest degree of cold, and the most powerful pressure, &c. Many of the pieces of apparatus also are extremely fragile, and require the most delicate management, and in many of the experiments the effects produced are so excessively minute, that only by the aid of the most perfectly constructed and complex arrangement of apparatus, used under the most favourable conditions, can they be detected at all. In microscopic manipulation we require to adjust the lenses to the exact focal distance, in order to see the lines on test-objects. In using a spectroscope we have to exclude extraneous light, and to separate as far as possible the compound lines without too much enfeebling them, and so on.

Descriptions of the methods of preparing and using all the different substances, the modes of making particular experiments, the manipulation and precautions requisite in using different kinds of apparatus, &c., may be found in all the textbooks of physics and chemistry. How to work with a microscope requires a treatise to itself, similarly also with a spectroscope and with the instruments and materials of each particular science and art. Some books are entirely devoted to descriptions of the modes of manipulation; for instance, amongst others may be mentioned Faraday's 'Chemical Manipulation,' Williams's 'Hand-book of Chemical Manipulation,' Kohlrausch's 'Physical Measurements,' Latimer Clark's 'Electric Measurement,' &c.

CHAPTER XXXIV

Observation of phenomena. Use of the senses in scientific research

THE term 'observation' is usually applied to the direction of our senses and perceptive powers to objects or phenomena, especially external ones. In order to observe a thing consciously we usually require to employ both our senses and perceptive abilities; thus we may be looking at an object, and although the rays of light proceeding from it strike our eye and excite the optic nerve, unless our attention is directed to it we do not perceive it.

Observation differs from experiment. By experiment we evolve facts, by observation we find them. Neither experiments nor other phenomena of nature impart to us new knowledge unless we observe their effects, because it is in the very act of observing that knowledge is acquired by the mind. In observing, we simply notice and record the conditions and phenomena presented to us, and it matters not whether they are those of prepared experiments, or those of phenomena over which we have no control.

Qualitative observation is very important. The discovery of a fact, however strongly it may have been anticipated or predicted, is not completely or really made until it has been actually observed. Accurate quantitative observation is also important; Tycho Brahè made an immense number of most exact records of the positions of the heavenly bodies with the aid of the best of astronomical instruments, and these records afterwards became the foundation of Kepler's well-known laws, and of modern astronomy.

Observation is closely connected with manipulation in making an experiment. Not only must the investigator manipulate so as to produce the desired effect, but he must know how to observe it when it does occur, also know what to expect, where to look for it, and what precautions to take in order to best perceive it. In some cases, before we actually make an experiment we write out a list of all the effects that we can imagine to be likely to appear, and this is a very good plan, because it points out to us what to look for. A logical habit of thought helps us in forming such a list, because it aids us in defining clearly the greatest possible number of alternatives.

During the progress of an experiment, as our perceptive faculties are very limited in delicacy, all the senses of the investigator must be on the alert in order to detect unexpected effects, as well as to observe those which are anticipated, or for which the experiment was purposely made, otherwise small and disguised effects will often pass unnoticed, and we must not forget that small effects, if they are unusual ones, are often the signs of great general phenomena. When Faraday first discovered magneto-electric action, the effects he obtained were so small that he could scarcely perceive them.

During the experiment also, all the results, and all the alterations made, and their effects, should be faithfully and truly recorded; and we must have no preconceived erroneous ideas of what to expect. In recording them we must be in the highest degree unprejudiced and exact, and neither add to nor detract from the truth; for nature will be sure to detect and expose any 'cooking of the results.' We must also endeavour to notice every essential particular, and make an accurate and complete record of them at the time, or at least before they have become at all dim in the memory. In taking notes of experiments, it is of but little use to record minor facts if the more important ones are omitted; the investigator should therefore be previously qualified to appreciate in some degree the importance of the more essential points. Generally also, as soon as an experiment, or a few experiments, have been made, conclusions should be drawn from the results, and suggestions noted, because the particulars are then fresh in the mind, and modifications can then be conveniently effected. Effects which cannot then be understood may be left for subsequent and more serious consideration. By a single observation we usually add a single new fact to our knowledge; but if that fact is an important one, like Oersted's discovery of electro-magnetism, it may implicitly contain a multitude of others. In some cases, persons who are skilled in routine observation only make more correct observations of the particular kind than those who know more of the subject generally, because their minds are less occupied by particular views.

Observations are usually concrete ideas, and often consist of a great many simple conceptions combined into one. When we merely observe the phenomena of nature we obtain ideas of facts (of different degrees of complexity), between which we may at any time make comparisons, from these comparisons draw inferences, and form decisions; and oftentimes we do so at once. In making comparisons, or drawing inferences, we also observe; in the former case we observe the facts or ideas we wish to compare, and in the latter, the propositions which we also desire to confront.

We acquire all the basis of our knowledge either through our external senses or by means of internal sensation and perception; and as we could not observe unless there existed something to be noticed, so the endless phenomena of nature, including those of our bodies and minds, are the original source of excitement of our observing power. The two general conditions of observation are nature and mind, *i.e.* the phenomena to be observed, and the mental power to observe them.

Observation is essentially active, and may be primarily divided into sensation and perception; both of which include mental activity. Every feeling occupies, to a greater or less degree, the entire mind; and perception may be viewed as a mental capacity of feeling impressions made upon the brain.

The senses may be enumerated as consisting of those of organic life in our physical frame, muscular feeling, touch, taste, smell, hearing, and sight. It is a peculiarity of the nerves of each particular sense, that an excitement of them produces only the kind of cerebral impression belonging to that sense. Thus excitement of the optic nerve, whether produced by light, mechanical or chemical stimulus, or by disease, produces only the impression of light, while the sensation of sound only is produced by the most diverse causes exciting the nerves of hearing. And as the nerves of the senses collectively ramify through the whole of our physical frame, and are all of them liable to be irritated by functional change or disease, it is obvious that when they are so affected we are subject to all kinds of sensations and perceptions, and are apt to make corresponding erroneous observations, unless we take proper precautions, and carefully examine the circumstances. As I have already indicated, in Chapter IX., some of the chief sources of erroneous observation, and it is beyond the scope of this book to enter into the details of correct observation in particular sciences, I must refer the reader to books on special sciences for information; and as a knowledge of the peculiarities, variations, and deceptions of each of the senses would assist a young scientific investigator to avoid making false or defective observations, I may recommend a perusal of a description of them contained in Professor Bain's work on 'The Senses and the Intellect.'

The power of observation differs not only in kind, but also in degree; some of our acts of observation are so feeble as only to feebly excite our consciousness, others excite it powerfully. Attention may be considered a high degree of volitional observation, and is a conscious mental effort to observe a sensation, perception, or idea, already present to the mind. The highest degree of observational action is both conscious and volitional. The feebler degrees of observation belong to the subject of 'unconscious cerebration.'

All our powers of observation are limited;[2] and the limits of their action depend both upon our inherited and acquired ability. 'We can hear from 20 to 73,000 sonorous vibrations per second, or an extreme range of between 9 and 10 octaves. The cry of a bat is too acute for some persons to hear';[3] and a practised ear can distinguish 1209 from 1210 simultaneous vibrations per second, of two tuning-forks. Professor Barrett has shown that sounds which are inaudible to us may be detected by means of a sensitive-flame. Various facts in natural history render it probable that some animals, and especially insects, possess either senses which we do not, or degrees of acuteness of sense exceeding ours. Izaac Taylor (I think it was) remarked that if our minds were not defended by the body from the influences of external nature around us, our lives would be rendered miserable by the acuteness of our sensations. It is not improbable that by administering certain gases or vapours to ourselves, and thus increasing the consciousness of a particular sense whilst diminishing that of others, extremely feeble sounds, odours, flavours, &c., might be detected which could not otherwise be perceived.

The power of observation is also liable to be interfered with, diminished, or prevented, by various causes. Simultaneous observations of diverse phenomena can only rarely be made; the observation of one thing, either by preoccupying the mind, or by exercising a more powerful attraction upon it, usually prevents our simultaneously observing another. Conflicting sensations are very effectual in preventing observation, because they are a powerful source of other ideas. As the mind cannot be simultaneously occupied by two ideas, one of which is so vivid as to occupy the whole of its attention, so an idea arising from a strong sensation usually shuts out all power of observing others.

That which one man misses in science, another perceives. This is often the case; for example, I discovered the sudden elongation of an iron wire at a particular temperature whilst under longitudinal strain during the act of cooling from a red heat; but Professor Barrett by repeating the experiment in a darkened room made an additional observation which I had missed, viz., that at the moment of elongation the wire suddenly evolved heat, and exhibited a visible and conspicuous momentary glow of redness.

As the great bulk of material phenomena are extremely minute, and our faculties of observation are exceedingly feeble, we observe only a small proportion of the effects, and those only the larger ones, which actually occur. Non-observation, therefore, does not prove non-existence, except in those cases in which it is certain that the objects if present would have been observed. As our faculties are powerless to appreciate perfect accuracy, and scientific instruments are more or less imperfect, the results we observe are always probable or approximate only. Many phenomena are also accompanied by others which cannot be separated. For example, current electricity is always accompanied by magnetism; chemical affinity is nearly always attended by evolution of heat; the action of gravity is a concomitant of every phenomenon; and so on. Some phenomena also are masked by others; for instance, positive and negative electricity often disguise each other. Sometimes we fail to observe, because we have not used the proper means to produce the greatest degree of effect, or we have not obtained a sufficiently conspicuous instance. In other cases where the effect is observable we have not employed the correct method of observation. It must also not be forgotten that, whilst making observations, we are very apt to be deceived by our senses; a large proportion of what we seem to observe, we do not really observe, but infer. False observations, stated as facts, are often more injurious to science than false theories; the latter can usually be disproved, but the former sometimes cannot, because the only opportunity of doing so may have passed away.

There are two general conditions favourable to the detection of minute substances or actions: first, by increasing the magnitude or intensity of the thing to be observed; and second, by the employment of more delicate tests or means of observing. Thus—1. 'An intense light will enable a smaller object to be seen. 2. A white picture can be seen smaller than a blue. 3. A line can be seen better than a point of the same diameter. The smallest angle for a round body is 20″; a thread-like object is discernible under an angle of 3″; a glancing wire can impress the eye at an angle of ⅕″.'[4] In the subject of chemistry, detection of substances in very minute amounts is usually facilitated by causing them to produce precipitates, or coloured solutions; and in other sciences, our powers of observation are immensely extended by the aid of telescopes, microscopes, spectroscopes, and a great variety of other instruments, specially designed for the purpose, also by the employment of special methods of manipulation. The particular kind of test, instrument, or mode of manipulation, is usually different with each different force, and every different class of phenomena. Descriptions of these, and of the various ways in which we may be deceived whilst observing different phenomena, are to be found in treatises on the various sciences. A large amount of in formation respecting the conditions and methods of observing and measuring phenomena relating to astronomy, magnetism, hydrography, tides, geography, geology, earthquakes, mineralogy, meteorology, winds, botany, ethnology, medicine, statistics, &c., is contained in the 'Admiralty Manual of Scientific Inquiry,' by Sir J. F. W. Herschel, Bart., and is well worthy of study by a young experimentalist. Many valuable observations also of a general character on the avoidance and elimination of error in physical observations, are contained in Jevons's 'Principles of Science,' chapter xv.

As it is an essential condition of successful research, that the investigator should possess clear ideas of his subject, every person who undertakes a research should be able clearly to describe his observations and conclusions in writing, and there is not, I think, an instance known in modern times of a valuable discovery in science having been lost through a want of power of description or exposition of his observations by a discoverer.

Sometimes one man observes and another generalizes; for instance, from the observations made by Tycho respecting the planet Mars, Kepler educed the path of that planet; and in a similar manner discovered that all the planets move in ellipses. Dalton based his atomic theory of chemistry upon the results observed by many preceding chemists.

The cause of original research is greatly aided by the labours of numerous scientific observers, who are continually taking observations and making notes respecting the winds, tides, earthquakes, magnetic changes, the amounts of daylight and of rainfall, and various other meteorological changes; and the time is fast approaching when nations must co-operate in promoting the collection of observations in other sciences, as they already have to some extent in magnetism, astronomy, meteorology, &c.

PART IV

Actual working in original research

Chapter XXXVIII

SELECTION OF A SUBJECT OF INVESTIGATION

> One science only will one genius fit,
> So vast is art, so narrow human wit;
> Not only bounded to peculiar arts,
> But oft in those confined to single parts.
> Pope, 'Essay on Criticism.'

Any man who wishes to discover new truth must be content usually to confine his search to one subject at a time. The selection of a good subject of examination is a difficult problem; the difficulty usually arises not from scarcity of subjects, but from the impracticability of determining which is the most suitable one. An investigator cannot, to any great degree, pick and choose discoveries, but must, to a large extent, be content to accept those he can find. In the selection of a subject of research he has to consider what subjects are intrinsically important, and that is often a difficult question.[5] In consequence of our very imperfect power of prediction we are very apt to value wrongly the importance of an unmade research; some of the greatest truths have been ultimately disclosed by investigating what previously appeared to be most trivial phenomena. The discovery of static electricity is said to have arisen from the circumstance that a bit of amber, by being rubbed, acquired the property of attracting a feather. That of magnetism was probably equally simple, and is said to have been first observed in a piece of loadstone. That of voltaic electricity also arose from an apparently trivial circumstance, which has already been described in this book. There are, however, cases where the investigator knows beforehand, with considerable certainty, from the nature of his proposed question and the conditions of his experiment, that if the hoped-for positive results are obtained, they must necessarily be important. The successful search by Faraday for magneto-electric induction, and his unsuccessful one for an experimental connection between gravity and the other physical forces, were instances of this kind. And there are other cases where the investigator knows beforehand that he is nearly certain to produce some new results, but is unable to foretell what they will be. In the first of these cases he wishes to know what conditions will render evident a particular new and important effect; and in the second, what effects will result from a particular cause or class of circumstances.

The investigator, in selecting his subject, has also to consider whether the conditions of the proposed research are sufficiently ripe, and that is another very knotty point. Even Newton himself could not discover the universal action of gravity when he first attempted to do so, because he made his first endeavour before the conditions were ready, and that could not have been ascertained without a trial. The ripening of a subject for research is affected by the length of time which has elapsed since the subject was last examined. The longer period of time a particular branch of science has been neglected, the more ready usually it is for a new investigation, because collateral branches of science have advanced and left it behind; several portions of inorganic chemistry, that of the fluorides for example, have been neglected, and are in this condition. Some investigators avoid examining a subject which has been recently and extensively investigated; it is, nevertheless, evident that however recently a subject has been examined, if from any cause it is likely to yield new results, it is in a fit state for further investigation. Great discoveries ripen very slowly. Now that the great discovery of spectrum analysis is a well-known truth, we can easily perceive that the conditions of it were maturing by means of successive researches, from the year 1802, when Wollaston detected lines in the solar spectrum, and the year 1815, when Fraunhofer measured their positions, until 1859, when Kirchoff and Bunsen discovered the composition of the sun's atmosphere, and thus suddenly made known its greatness.

Another very uncertain point in selecting a subject is the degree of probability of success in making the research. In making new experiments, all kinds of obstacles and new effects arise which we cannot foresee, and from that and other causes, such as our limited means of detecting effects, we rarely obtain from such experiments the expected results. An investigator has not unfrequently to advance some distance into a preliminary research before he can determine whether or not he has really a definite and new question upon which to work; and he often finds, after much expenditure of time and trouble, that the supposed new phenomenon is only an old one in a disguised form, and that his hypotheses therefore respecting it are all wrong.

Every investigator also soon rejects questions which are beyond his powers. We are often much better able to judge of the degree of importance of a proposed research than of our ability to carry it out. Our ability to discover is, in most cases, inversely proportional to the degree of intrinsic importance of the desired issue. Many researches might easily be selected which we know beforehand must yield positive, new, and distinct effects, but such investigations are nearly always of a comparatively unimportant kind, because the results of them are generally only additional instances of a similar class to some already known; for instance, most of our tables of constants might be largely extended. It would be hardly possible to suggest a research which would be certain to yield a perfectly anomalous phenomenon. Perhaps the most probable way to suggest a research which would be likely to yield one would be to assume an hypothesis that any substance which is known to behave in an anomalous manner with regard to one force would also behave similarly with regard to another force.

The embarrassment of selection is usually caused partly by a desire to obtain valuable discoveries at little trouble and expense, and partly by our being so little able to predict successfully new important effects. We know so little about what is termed the 'internal resistance' of substances, which is believed to determine largely the special effects of different forces upon them, or of the molecular structures and motions which form essential portions of nearly all physical and chemical phenomena, that the selection of a subject of research is, to some extent, in many cases a 'leap in the dark.' Those experiments, however, are usually rejected which appear extremely uncertain, unless they can easily be made, and would, if successful, probably yield important results.

A very good plan, and one which I have adopted on various occasions with perfect success, has been to devise an arrangement, and select a research, in which matter or its forces was placed under new conditions, and trust implicitly to the general truth that every new arrangement of matter or force must produce new results.

In other cases the difficulty is usually overcome by selecting from a stock of hypothetical suggestions and questions those which appear to have the greatest degrees of importance, probability, and ripeness, and adopt the most suitable one. Particular researches are sometimes selected, because they are less expensive. In some cases, however, a research is not, strictly speaking, selected at all, but the investigator is led on from a previous inquiry to another by questions which arise at the time, and, having all the materials and apparatus more ready at hand than if he commenced an entirely different subject, he prefers the former; for instance, Faraday appears to have been led on to his discovery of the important law of definite electro-chemical action from his immediately preceding experiments on electrolysis and electric conduction.[6]

Whilst one scientific man expends his time upon comparatively trifling matters, another slowly and persistently works out a great idea. Most of the ablest of discoverers appear to have acted, to a large extent, upon the plan of exerting the greater part of their strength upon important subjects, and have selected those questions and experiments which, if they can be solved, or can be made to yield a positive result at all, must yield one of importance; such, for instance, as definite experiments to test the existence of a new relation between two forces. Oersted acted upon this plan; he asked the question, 'Are electricity and magnetism really related to each other?' and ultimately discovered electro-magnetism; and Faraday in particular employed it, and found, after many trials, magneto-electric induction, and the relation of magnetism to light. Andrews also appears to have acted upon it in his researches on the continuity of the liquid and vaporous states of matter.

Chapter XXXIX

OUTLINE OF A MODE OF CONDUCTING AN ORIGINAL RESEARCH

As it may be of service to young experimentalists, I give the following condensed outline of the chief steps to be taken in carrying out one of the commonest forms of original qualitative research in physics or chemistry.

As discoveries are originated in a great variety of ways, it is assumed that in this case the investigator is already acquainted with some phenomenon or mode of working, original or otherwise, which he believes may, by his taking the requisite trouble, yield some new results.

The first step to be taken is to prepare the necessary substances and apparatus; and if it is a chemical research, to ensure at the outset the highest attainable degree of purity of the substances. If apparatus is required, it is necessary to consider carefully first its general plan (aided by a sketch) and its proper magnitude, then each part of it in succession, so as to secure what appear to be the essential conditions of its action, and to exclude those which would probably prevent or diminish the expected effect; and then to arrange the most delicate means of detecting, and, if necessary, also measuring the effect.

These being provided, a few preliminary experiments may now be made, and as many conclusions drawn from the results of them, and from comparisons of the results with each other, as are proved by the evidence. During these experiments as many hypotheses as possible should be raised by studying the results and conclusions, and notes be kept of all the results, conclusions, remarks, and hypotheses for future reference.

The sources of error and interference should now be excluded, and the phenomenon be obtained, or the method of working be arrived at, in a pure or perfect state and free from unessential conditions, as soon as possible. This is effected by selecting those hypotheses which bear upon the particular points, and testing them by additional experiments and observations, in which each condition is excluded, one only at a time; and this usually requires very varied experiments and considerable thought and trouble.

Having at length obtained, by means of these additional experiments, an approximately perfect form of the phenomenon or method of working, the next step is to make a systematic and comparatively exhaustive examination of each condition of the phenomenon or method, and also to make the experiment with, or employ the method upon, every suitable substance. By making the experiment with every possible substance, the greatest number of conspicuous and exceptional cases will be included; exhaustive researches also generally yield the greatest discoveries, because it is the exceptional cases of the exceptional ones which disclose in the greatest degree the most hidden conditions. During this examination every logical conclusion possible should be drawn from the results,[1] and hypotheses raised upon these conclusions, at frequent and suitable intervals of time, while notes should also be continually made of the results, conclusions, remarks, and hypotheses.

After having made every possible experiment, the utmost amount of new knowledge we are able to obtain should be extracted from the results. This is effected by classifying and combining them in every conceivable way, and stating all the observed uniformities of each class in the form of general conclusions. If, during the process of classification, we meet with exceptional instances, we must, in order to harmonise the general conclusion with them, infer a still more general conclusion which will include both the ordinary instances and the exceptional ones. It is in this way that the apparent truth is often shown to be different from the real, and the results of superficial examination to be opposite to those of deeper research; the apparent cause only of a phenomenon being disclosed by usual instances, and the real and deeper cause by exceptional ones.

If we wish to extract more completely the truths implicitly contained in the results, we must subject the general conclusions themselves to every possible variety of combination and permutation; and to obtain a still further amount, we must add to the conclusions truths from other branches of science, and re-perform the same processes.[2]

The discovery of the general principle or law which pervades and regulates all the results usually completes the research. Sometimes, however, in addition to this, a perfect specimen of the apparatus is constructed for the purposes of illustration; also a practical form of it, for the purpose of showing that it is probably capable of application to some immediately useful purpose. This latter act, however, is one not of scientific discovery, but of invention, and is a step towards an artistic, manufacturing, or commercial application of the discovery.

In many cases, the last performance in an original research, viz., the drawing of general conclusions, and explaining the results, are much more complex and difficult than would be inferred from the foregoing remarks, because physical and chemical phenomena are often results of a combination of causes which cannot be separated, and are attended by a number of concomitant effects and circumstances which we cannot exclude, and which render it difficult to draw correct conclusions, and also to prove and explain the true relations of the phenomenon. Indeed, the complexity of nature far surpasses, in nearly all cases, man's usual ideas of it; the simplest phenomena of matter present almost infinitely more for us to explain than we can even imagine. The modes of discovering causes, coincidences, and other relations of phenomena, are described in Chapters XLVI., XLVII., and XLVIII.

In an original research, we encounter difficulties at every step, and we must, if possible, overcome them. The fundamental rule for overcoming a difficulty in research is to attack it in detail, *i.e.*, to divide it to the fullest extent, and analyse and treat each part separately. Thus when we wish to know the exact chemical composition of a complex substance, we dissolve one hundred parts of it, and then precipitate or otherwise separate its ingredients in several collections, known as the hydrochloric acid group, sulphuretted-hydrogen group, sulphide of ammonium and carbonate of ammonium groups, the alkaline earth and the alkaline groups; and we then subdivide each of these collections into several smaller ones, and so on, until we have individually isolated each of the elementary bodies present in the original substance, and can subdivide them in that manner no further. We then ascertain if the sum of the weights of the elementary substances found exactly equals that of the original body. If it does, the analysis is complete; but if it does not, we have either omitted some substance present, introduced a foreign body, or committed an error of manipulation, observation, or calculation; and we proceed at once to correct the mistake.

In the examination of a physical phenomenon, we proceed in a substantially similar manner, *i.e.*, we divide the conditions of it in every possible way, and into the smallest distinct elements, and examine each portion, not only in a qualitative manner, but also quantitatively, as far as the circumstances of the case will admit. The quantitative analysis of a physical phenomenon, however, can often be only incompletely performed. In the analysis of such a phenomenon, instead of dividing its conditions first into groups, and then each group into its individual distinct elements, as in chemical analysis, we usually separate one only of its conditions at a time, leaving each time all the remaining ones together; and we then sometimes reduce the complexity of the phenomenon by an element at a time, or subdivide the phenomenon in other ways, according to the circumstances of the particular case. In every case, however, we reduce the phenomenon to its purest and simplest state, by separating from it all its unnecessary conditions, as soon as we can, in order to fit it for the actual analysis.

Chapter XL

ADVANTAGES OF VARIETY OF EXPERIMENTS

A VARIETY of well-devised experiments is of the highest importance, and is much more likely to yield valuable discoveries than any mere number of similar ones, especially if the questions involved in the experiments are some of them intrinsically momentous. It also supplies us with results and data by means of which we can, by processes of comparison and inference, determine the causes, coincidences, and other abstruse relations of phenomena. Variety of experiments is also the chief means by which the indifferent and unessential circumstances, and those which interfere with a phenomenon we are investigating in a new research, are quickly disclosed and detected. If the conditions of an experiment are not sufficiently varied, we can hardly be sure that unsuspected influences have not affected the results; nor can we be certain that our interpretation of the results is correct. A uniform method of working may conceal a uniform error. In every experiment there are always many ways of failing, and usually only one way of succeeding; and no precise rule can be laid down, either for preventing errors in experiments or for excluding those errors which have been detected, because the methods of preventing or excluding them differ somewhat in every different case. Failure in experiments often arises from want of proper materials, or proper proportions of them, and in a great variety of other ways. 'Although the truth is not always arrived at by the first experiment, that is not the case because the first idea of the experiment is not very often quite adequate to obtain the truth; but it may sometimes happen because the materials and means which are used to carry it out practically are not adapted to that purpose; and although these experiments cannot contaminate the purity of the theoretical speculations, they are nevertheless unfitted to second them, on account of the materials employed.'[7]

In an investigation, it is usual to exclude the largest errors first, because it is not of much use to exclude small ones if large ones remain; and it is highly important to reduce the mode of experimenting to its purest and simplest form as quickly as possible, because until that point is attained the results of the experiments are more or less untrustworthy, and, no matter what time or money they may have cost, have to be wholly or partially discarded.[8] The mode of experimenting cannot, however, be reduced to its purest form until such a variety of experiments have been made as to disclose the existence of all the interfering circumstances. Some interfering circumstances are very treacherous sources of error, because they are unsuspected; such are usually disclosed by a want of uniformity in the results, even when the conditions are apparently similar; and this want of uniformity is produced by the previously unsuspected cause or causes operating according to a different law of variation to that of the cause of the pure phenomenon. As an example of this kind; I was investigating the phenomenon of electric currents produced by two unequally heated pieces of platinum in liquids which did not corrode them, and employed an apparatus which was rendered water-tight by means of washers of pure indiarubber. After making many experiments, I found that, although the indiarubber had no chemical action on platinum, it chemically altered some of the liquids, and thus affected the electric currents, though not to such a degree as seriously to influence the results. I had, however, to abandon the use of an expensive apparatus and devise and employ a different one, devoid of indiarubber in contact with the liquids, and also to repeat with it most of the former experiments, in order to compare the new results with the old ones, and thus determine the extent of the interference. Also, in modifying an experiment, or the method of working, in a research, whether the modification is intended for the purpose of saving time or expense, removing an inconvenience, or for other reasons, it is necessary to consider previously whether in so doing we may not introduce an interfering condition.

As neither our apparatus, materials, or modes of manipulation are perfect, the results obtained by means of experiments must always be contaminated with more or less error, and therefore extremely exact results are usually suspicious.

Chapter XLI

ADVANTAGES OF NUMBER OF EXPERIMENTS

In research it is often a good plan to settle off-hand the questions which arise at the moment, and thus clear one's mind of doubts as they arise; but if this practice is carried too far, as it very easily may be, it is certain to lead the investigator away from his subject into a multiplicity of researches which he will never be able to complete. In searching for new truths, instead of waiting until we can predict with certainty, which in most cases would be to wait a very long time, it is frequently better to make experiments at once, because the making of a number of different experiments, if varied and important ones are included, is one of the great and prime conditions of discovery. Some of the discoveries made by Dr. Priestley are examples of this. The experiments must, however, be new and well-conducted;

if they are not new, the results will not be novel; and if not well-conducted in all essential respects, the results will be erroneous, and more or less misleading. A single well-chosen and properly carried out experiment is far more valuable than a multitude of imperfect ones; if an experiment is well made, a repetition of it is almost unnecessary, because it would add but little to the certainty of the results. It is not the number of experiments alone which lead to great discoveries; Potts of Halberstadt, Professor of Chemistry in Berlin, and a favourite of the King of Prussia, is said to have performed 30,000 experiments in six years, for the purpose of ascertaining the ingredients and process employed in making Dresden porcelain, but discovered no important principle in science. He, however, by these experiments laid the foundation of blowpipe analysis; he also discovered in the year 1740 that oxide of manganese was a substance different from oxide of iron, and in 1746 showed that silica was distinct from other earths.

When once a phenomenon has been divested of all sources of interference, and obtained in its purest and simplest form, a large number of experiments is of great importance, because if we repeat an experiment with every possible variety of substance, we are sure to include the most feeble and the most conspicuous instances of the phenomenon, and also the exceptional and contradictory cases, if there are any. The discovery of extreme and exceptional instances is often of great value, because it enables us to draw important conclusions, and by the further investigation of such instances we sometimes detect an entirely new class of phenomena which an ordinary instance would not enable us to suggest. But until the particular phenomenon has been obtained in its purest and simplest form, a multiplicity of experiments yield such a mass of complicated and conflicting results that even the clearest and most discerning intellect is utterly unable to disentangle and explain them. The phenomena of an experiment, even in its simplest condition, are usually related to or inseparable from so many other phenomena,[1] especially in the biological sciences, that it is extremely difficult to classify and harmonise all the results, some usually remaining unexplained even in a finished investigation. These circumstances indicate the comparatively small value of conclusions derived from the results of crude experiments.

The full explanation of a phenomenon requires both variety and number of experiments; variety, to enable us to exclude interferences and discover causes; and number, to detect laws, general principles, and extreme and conspicuous instances.

Chapter XLII

IMPORTANCE OF MEASUREMENTS

'God has made everything by weight and measure.'

NEXT to qualitative truths themselves, their quantitative relations form the essence and basis of science. Scarcely a scientific research can be made without the assistance of measurements. During the early periods of science, the attention of mankind was limited chiefly to facts, and did not largely extend to quantitative relations; measurements were then less frequently made, and those which were made were comparatively crude and deficient in accuracy. Even now, in a research, unless the question to be solved is in itself primarily a quantitative one, measurements are not made until the existence of the phenomena themselves are ascertained. But in all the sciences, we have now more or less passed the logical or qualitative stage, and have entered, to a greater or less extent, into the sphere of exact quantitative research. Sir William Thomson, in his inaugural address, delivered in 1871, to the members of the British Association, says, 'Accurate and minute measurement seems to the non-scientific imagination a less lofty and dignified work than looking for something new. But nearly all the grandest discoveries of science have been but the rewards of accurate measurement and patient, long-continued labour in the minute sifting of numerical results. The popular idea of Newton's grand discovery is, that the theory of gravitation flashed upon his mind, and so the discovery was made. It was by a long train of mathematical calculation, founded on results accumulated through prodigious toil of practical astronomers, that Newton first demonstrated the forces urging the planets towards the sun, determined the magnitude of those forces, and discovered that a force following the same law of variation with distance urges the moon towards the earth. Then, first, we may suppose, came to him the idea of the *universality of gravitation*; but when he attempted to compare the magnitude of the earth's surface, he did not find the agreement which the law he was discovering required. Not for years after would he publish his discovery as made. It is recounted that, being present at a meeting of the Royal Society, he heard a paper read, describing geodesic measurement by Picard, which led to serious correction of the previously accepted estimate of the earth's radius. This was what Newton wanted. He went home with the result, and commenced his calculations, but felt so agitated that he handed over the arithmetical work to a friend; then (and not when, sitting in a garden, he saw an apple fall) did he ascertain that gravitation keeps the moon in her orbit. Faraday's discovery of specific inductive capacity which inaugurated the new philosophy, tending to discard action at a distance, was the result of minute and accurate measurement of electric forces. Joule's discovery of thermo-dynamic law, through the regions of electro-chemistry, electro-magnetism, and elasticity of gases, was based on a delicacy of thermometry which seemed impossible to some of the most distinguished chemists of the day. Andrew's discovery of the continuity between the gaseous and liquid states was worked out by many years of laborious and minute measurement of phenomena scarcely sensible to the naked eye.'[9]

'Here, then, we have a very full recognition of the importance of accurate measurement, by one who has a perfect right to speak authoritatively on such a subject. It may indeed be maintained that no accurate knowledge of anything or any law in nature is possible, unless we possess a faculty of referring our results to some unit of measure, and thus it might truly be said—"*to know is to measure*." '[10]

After having ascertained the existence of a fact or principle in science, the next most important step is to determine its amount under various conditions of time, space, direction, &c., and its quantitative relations to its causes and effects; to other facts and principles, &c. Often by the aid of measurements we are enabled to determine the nature, causes, and conditions of phenomena. One of the commonest methods of discovering whether two different phenomena are connected together is to ascertain whether they simultaneously vary in amount; if one varies in the same proportion as the other which accompanies it, the two are probably related to each other, either as cause and effect, or as coincident effects of the same cause. Each force can only act in accordance with its own laws; the quantitative relations and properties of forces, therefore, often enable us to detect and distinguish the action of those forces in any new phenomena which we have discovered, and to ascertain also the extent to which such forces exist and operate in any given case; for instance, the property of chemical force, of causing dissimilar substances to unite together only in definite proportions by weight, and produce a third homogeneous substance of widely different apparent properties, enables us to detect that force in any new case of chemical change that occurs.

Quantitative methods have been and are almost inexhaustible sources of new scientific knowledge, and a multitude of discoveries, many of which are of the highest value, might be described, which have resulted from the employment of processes of accurate measurement. By means of them Galileo discovered the laws which regulate the descent of falling bodies, and found that, when the time of descent was doubled or tripled, the space traversed became four or nine times greater, and therefore that the spaces fallen through were proportional to the squares of the times of descent; and a knowledge of this law, that the earth exercised the property of attracting bodies, with a power which varied inversely as the square of the distance, largely enabled Newton to discover the universal action of gravity, by showing that the law agreed with the motions of the heavenly bodies. Great progress in chemical science has been due to the introduction by Lavoisier of the use of the balance; and the errors in Stahl's theory of phlogiston were chiefly discovered by the aid of that instrument; the chemically combining proportions of substances were found by means of accurate weighing; and the use of the goniometer in measuring the angles of crystals resulted in various geometrical discoveries in the science of crystallography.

One of the grandest proofs of the value of measurements was the discovery of the equivalence of all the physical forces and chemical elements, and of their indestructibility; and another was that of the system of atomic weights in chemistry. The laws of action of the forces of heat, electricity, magnetism, have all been ascertained by the combined application of qualitative and quantitative methods; and the whole of the sciences are penetrated in all directions by laws requiring for their discovery both arithmetical as well as logical processes.

Quantitative measurements also greatly aid us in discovering residues of substances, forces, or effects. It is clear that as neither matter nor force can be destroyed by human agency, from one hundred parts of a mixture of substances taken for analysis we ought to be able to obtain neither more nor less than one hundred; and, similarly, if we expend one hundred parts of a given force, we ought to be able to obtain as a result, in the form of other forces, a total which is equivalent to neither more nor less than the whole amount of power expended.

Some phenomena (such as those of small periodic changes of terrestrial magnetism) are so hidden by larger phenomena of a similar kind, that they could hardly be discovered at all except by the aid of long-continued series of systematic measurements, and taking the average of numerous results. The average strength and direction of the wind, the mean pressure of the air, the true sea-level, and many other facts, are arrived at in this way.

Some quantitative processes and contrivances save much labour in research. It has been said that the invention of logarithms doubled the life of astronomers, and that that of the differential calculus was as great an aid to scientific discovery as that of the steam-engine was to the mechanical arts.

In consequence of the great flood of light which is thrown upon every scientific truth, by a knowledge of its quantitative relations, we should in all cases of original research make measurements of the phenomena, and all their attendant conditions and circumstances, whenever practicable, and especially of the most important ones. As the phenomena to be measured are often very minute, and their magnitudes cannot be estimated by means of our senses, and as the instruments we employ to measure them always admit of some interferences and some degree of inaccuracy, it is of great importance, in nearly every case, to employ the most trustworthy and accurate instruments. Every new refinement in our methods and instruments of measurement is sooner or later employed for making additional discoveries. Instruments of precision have immensely aided discovery, and every investigator should therefore possess those adapted to his researches—such as accurate goniometers, barometers, balances, microscopes, spectroscopes, photometers, thermometers, electrometers, galvanometers, &c. Accurate measurement is also very important for the purpose of making units, standards, rules, measures, verniers, zero-points, maximums, minimums, speed-ratios, &c. &c.; and the use of such accurate units, standards, &c., is also of very great value. Less accurate instruments or measurements cannot be employed to verify the truthfulness of more accurate ones; it is of but little use to measure an effect unless we also measure its cause and conditions, nor to measure the cause or conditions unless we also measure the effect, nor to measure the one much more accurately than the other. Tables of terrestrial constants, obtained by daily measurement and record of the phenomena of light, heat, magnetism, and electricity, all over this globe, are also of immense value in leading to discoveries respecting this planet and other heavenly bodies, and their true functions in the universe.

In order to obtain accurate ideas of substances and forces, we require to know the exact degree of every quality and property which the substance or force exhibits. Nearly all the properties of substances and forces are of degree only, and not complete: for example, no substance is perfectly elastic, or an absolutely perfect conductor of heat or electricity; no one body is absolutely transparent or opaque to light or heat; all substances have different specific gravities, and even the solid, liquid, and gaseous states of matter merge into each other by insensible degrees; and all these require to be measured under varied conditions. Very few of the properties of forces or substances are entirely characteristic—nearly all of them are relative, and possessed in different degrees by several or many substances: for instance, the property of attraction is not a characteristic of magnetism—electricity also exhibits it; magnetism is not characteristic of iron, because cobalt and nickel are also magnetic; great specific gravity is not possessed by platinum alone—several other of the noble metals are nearly equally heavy, &c. &c. In the subject of chemical analysis we perhaps find the most perfect examples of characteristic properties of bodies, but even in this case a substance can rarely be distinguished by means of a single property alone, but nearly always only by several.

The special degrees of the different properties possessed by different bodies are much more frequently characteristic of the different substances than are the properties themselves. With regard to most of the properties of matter, each substance, in the pure state, usually possesses its own special or characteristic degree of each property under the same conditions of pressure, temperature, &c.: for instance, the specific gravity of hydrogen, being assumed to be equal to 1, that of nitrogen is 14, oxygen 16, chlorine 35½, &c.; the specific gravities of each of the pure metals is also a nearly definite and characteristic number.

Chemists partly detect and distinguish the various elementary substances, as well as their compounds, by means of the different degrees in which they possess particular properties, and they determine the amounts of those substances by means of their different atomic and molecular weights, and combining proportions. In qualitative analysis, although the degrees of the properties of substances are rarely actually measured, they are estimated by the eye, and employed as qualitative tests: for instance, a precipitate of chromate of lead is distinguished from one of chromate of baryta by its greater degree of yellowness; one of sulphate of barium from one of sulphate of calcium by its greater degree of insolubility in water; a salt of strontium from one of calcium by the greater degree of redness which it imparts to a flame, &c. &c.

In these, and a multitude of other ways, quantitative research, and mathematical, arithmetical, and geometrical methods, are of immense use in physical and chemical discovery. We know, for instance, *à priori*, from geometrical considerations, that a molecular chemistry of one dimension only cannot be true, and that any true molecular theory of physics or chemistry must admit of three dimensions.[11]

Millions upon millions of measurements remain to be made in every single science. All our tables of constants remain to be completed, and the old ones reconstructed, by the aid of more refined methods and instruments, and a very great number of new ones formed. Nearly all our tables of physical and chemical phenomena are limited to the ordinary range of atmospheric pressure and temperature, but we require additional ones for all pressures and all temperatures, and for all combinations of these. A probable result of the formation of such tables will be a profound alteration of our views of nature, because every single substance has more or less different properties at every different pressure and temperature.[12]

'Now with respect to accurate measurement, theory was left far behind by practice, and I need not to be reminded how very much more accurate were the measurements of resistance in the practical telegraphy of Dr. Werner Siemens and his brother than in any laboratory of theoretical science. When in the laboratory of theoretical science, it had not been discovered that the conductivity of different specimens of copper differed at all, in practical telegraphy workshops they were found to differ by from thirty to forty per cent. When differences amounting to so much were overlooked, when their very existence was not known to scientific electricians, the great founders of accurate measurements in telegraphy were establishing the standards of resistance accurate to one-tenth per cent. Dr. Werner Siemens and his brother were among the first to give accurate standards of resistance, and the very first to give an accurate system of units founded upon those standards.'[13]

Accuracy of measurement is often obtained by means of repetition, as in the pendulum; also by means of coincidence, as in the vernier. Descriptions of the numerous ways in which measurements have led to discoveries, and of the various methods of measuring substances, forces, conditions, and actions; of weighing, and taking the specific gravities of solids, liquids, and gases; of determining the coefficients of elasticity and of numerous other properties, the positions of spectral lines, the degrees of conduction-resistance to heat and electricity, also of paramagnetic and diamagnetic capacity, and a multitude of other phenomena, at all temperatures and pressures, form a portion of the subject of quantitative research, and may be found in the various text-books of the different sciences.[14]

Chapter XLIII

COMPLETION OF RESEARCHES

It is desirable that researches which have been commenced should be completed. Not unfrequently they are abandoned when only partly made. The difficulty of completing them arises partly from the immensity and complexity of nature. The most trifling fact, when exhaustively investigated, gives rise to many varied questions, each of which requires a separate research, and these separate researches in their turn give rise to others, so that it appears impossible to exhaust the subject, or make the first research complete, and thus the results of valuable investigations are in consequence often withheld from publication: this is a common case. Researches are also sometimes put aside in an incomplete state by the investigator for the purpose of pursuing new ones, which appear more important or more attractive.

A complete research should be exhaustive, because it often happens that if a research is not thorough, instances of an important or exceptional kind remain undiscovered, essential similarities or differences continue unnoticed, and the true explanations may remain unattained and a wrong one be assumed. Some scientific men, having made an experiment and observed its result, or observed some singular natural phenomenon, stop short, and fail to examine it further. Others make a partial investigation only, either making an insufficient variety of experiments, and thereby failing to exclude interferences; or an insufficient number, and thus missing extreme and exceptional instances.

Every part of an investigation, and even each single experiment, may however be viewed as a smaller research, complete as far as it goes, because we can compare and classify results, and draw analogies and inferences from them as we proceed, and also infer from each result and class of results the full amount of conclusion warranted by the evidence. Viewed from the opposite aspect, the most extensive research is never complete, and never can be until it extends to the utmost bounds of knowledge.

Notes

1 Address by Dr. Frankland. *Conferences:* Loan Collection, London, 1876, vol. ii. p. 11.
2 See Chapter XXIV.
3 Bain, *Senses and Intellect,* 2nd edit. p. 215.
4 Bain, *Senses and Intellect,* 3rd edit., p. 220.
5 See Chap. XIX.
6 *Life of Faraday,* by Dr. H. B. Jones, vol. ii. pp. 20–35.
7 Magalotte.
8 I believe that Mr. Bailey, in his experiments made to ascertain the mean density of the earth (commonly known as 'weighing the earth'), after several years of labour and great expense, discovered an error which vitiated all the results; and he therefore discarded the whole of them, and repeated the entire series of experiments over again, with the source of error excluded from them.
9 Address to the British Association, 1871.
10 Address by Dr. C. W. Siemens. Conferences, Special Loan Collection, London, 1876, p. 206.
11 See *Geometric Chemistry*. By Henry Wurtz, 1876, U.S.A.
12 See Chapter IV. p. 34.
13 Address by Sir W. Thomson. On Electrical Measurement. Conferences, Special Loan Collection, London, 1876, p. 247.
14 For illustrations of the great value of mathematics in scientific discovery, see Whewell's *Philosophy of the Inductive Sciences,* vol. i. pp. 150–156; Whewell's *History of Scientific Ideas,* vol. i. 3rd edition, chap. xiv. pp. 153–167; also G. C. Foster's Address to Physics Section of the British Association, 1877.

Editorial Headnote

Arthur Mason Worthington, *A First Course of Physical Laboratory Practice* (London: Longman, Green, and Co., 1890), 12–26

Arthur Mason Worthington, 1852–1916, known principally for the development of high-speed photography as a means of mechanical objectivity in the observation of splashes. This text is aimed at establishing the experimental method as a course of study for children at school. He makes numerous references to Clifton College, a Public School in Bristol where he was Science Master in 1877 and between 1880 and 1884.

16

A FIRST COURSE OF PHYSICAL LABORATORY PRACTICE

Arthur Mason Worthington

Source: London: Longman, Green, and Co., 1890, 12–26

THIS book, which is intended to supersede a little volume published a few years ago, is written with the object of encouraging the teaching of Physics in schools, especially by means of laboratory work, in which the pupil performs his own experiments. It embodies the Experimental Course which has been found suitable for *beginners* at Clifton College, where the author has had several years' experience in conducting large laboratory classes of as many as thirty pupils at a time.

In the very valuable report of the United States Government Bureau of Education, edited by Mr. C. K. Wead of the University of Michigan, and published in 1884, there occurs among the answers to a question as to the desirability and feasibility of laboratory work in schools, one which urges that it is objectionable "to put students to experimental work when unqualified for it, and with inadequate means. Thus habits of slovenly experimenting and inconsequent induction are formed, or the student is disgusted with the unsatisfactory nature of the whole thing. Especially is a loose way of experimenting with cheap apparatus, and obtaining only the remotest approximation in results—results which would not of themselves even suggest the principle or law—very much to be deprecated." I think it not unlikely that an objection of this kind may be felt by some of our own scientific men who may be without personal experience in teaching the elementary portions of the subject to young students, and therefore take this opportunity of saying a few words on the point.

Our experience at Clifton has shown that experiments can be selected which shall lead to quantitative results quite sufficiently accurate to suggest or confirm the correctness of the principle involved without taxing too much the patience or the intelligence of the young experimenter. And we have found that, so far from being disgusted with the unsatisfactory nature of the experiment, he is generally encouraged by feeling that an accurate result is within his reach if he takes sufficient pains. To be sure, his own standard of accuracy is a low one in comparison with that of a trained Physicist, but this is only a symptom of his undeveloped condition, and as he becomes more practised, his standard improves of itself. It is precisely thus that the standard of precision in all attainments grows gradually

DOI: 10.4324/9781003009337-22

with practice and increasing mental power. How many stages lie between the games of young children and the same games as played by men! Such a laboratory course as I advocate recognises the low standard of particular faculties, and gives them an early opportunity of natural growth. No doubt it would be prejudicial to keep a pupil very long at work in which the highest standard attainable was kept out of sight. For this reason this preliminary laboratory work at Clifton is confined to boys in the Middle School, *i.e.* of about the ages of fourteen to seventeen, after which point it either ceases or is exchanged for more exact work, modelled more or less on the teaching in the University laboratories.

But there is another aspect of this work to which I would call attention. It should not, I think, at first be the aim of the teacher to show his pupils how measures of Physical Constants may be *most accurately* made, so much as to give them a clear idea as to what such measures *mean*,—to give them, in fact, a thoroughly intelligent conception of what is to be aimed at. And for this reason the methods of measurement should at first be, if possible, such as most naturally suggest themselves to the mind of the learner, or such as are least encumbered with secondary considerations, and all subordinate "corrections" likely to obscure the main issue should at first be avoided, while at the same time the pupil should make the measurement himself.

In this respect I conceive that the teaching of younger should differ from that of older pupils who have already attained sufficient grasp and patience to face with pleasure the complicated considerations essential to the more accurate measurements, though, even with senior students, what may be called "instructive" rather than "accurate" methods of measurement may sometimes be selected with advantage. For among the ultimate ends of our science-teaching, to the great majority of pupils, who are not likely to become specialists, we must surely reckon as very important the interest in natural phenomena that results from mastering some principles of Nature, and the resources of intellectual enjoyment which such interest throws open in after life.

Of the educational value of the systematic, practical study of Physics, as here advocated, those who have watched it at Clifton, and have observed its popularity with the boys, have little doubt. It is undertaken there, like all the scientific teaching, not with a view of training up Physicists, but with the object of evoking in the boys a genuine and generous interest in natural phenomena, and of training them to habits of patient and conscientious study; and those of us who have devoted themselves more particularly to the Physical Sciences are confident that the serious interest thus early aroused in a large number is the best guarantee of future excellence in the few who may afterwards become specialists.

To pass to matters of detail:—I have found that it is desirable, when setting a young student to perform an experiment for himself, to give him something more than a bald description of what is to be done. It is necessary in many cases to point out clearly what the experiment is designed to teach, and often for this purpose to summarise information already acquired, so that the significance of the experiment and its results may be more fully perceived; and unless this be carefully done, the work loses much both of its value and interest. For this reason the experimental practice should always be accompanied or preceded by experimental lectures, or by a carefully written comment on the experiment that the student is himself to perform. And I have found that the former will not always supply the place of the latter, since it is not always possible to work the two systems of instruction hand in hand. Exigencies of school arrangement will sometimes prevent the experimental classes from consisting of quite the same individuals as the lecture classes, and it often happens that the lectures will get ahead of the experimental work, or the experiments of the lectures. Consequently the commentary that accompanies the experiments which the pupil is to perform should repeat (it may be in a condensed form) the information that is given in the lecture.

In the course, therefore, which follows, it will be found that the remarks by which the experiments are introduced or explained often assume the character of a written lecture; the object which was in my mind being to afford to the student all the requisite materials for teaching himself.

At Clifton the senior boys take home their rough notes of experiments performed, while the juniors write out the fair copy before leaving the laboratory—always in pencil, no ink being allowed in the room.

The boys work in pairs, an arrangement found on the whole quite satisfactory, and each pair works quite independently of the rest, some getting on much faster than others. This independence is an advantage, not only because it renders it unnecessary to keep apparatus enough for all to be doing the same thing at once, but also because it makes each boy feel that his progress is in his own hands.

The laboratory is a large well-lighted room, 28 ft. by 22 ft., fitted with suitable tables, cupboards, gas, and water. The nature of the fittings may be gathered from the cut on the opposite page, which shows a work-table for four boys (two pairs). Below each half of the table is a cupboard with a shelf containing the apparatus in most frequent use. For the sake of light, these cupboards are not separated by any internal partition. At the back is a long narrow cupboard running the whole length of the table, with a shelf and a door at each end; in this are kept the note-books or special apparatus in current use. On one table is seen the moveable staff for holding the balance, on the other, a burner connected with one of the pear-shaped gas heads. The top is of well-seasoned deal, 2 in. thick, with a square-cut edge, and projects 5 in. over the cupboards. The total length of the table is 6 ft., depth 3 ft., and height 2 ft. 10 in., and the two halves of the top are separated by a moveable strip of wood.

Near each half of the table is seen a water-tap and small sink, provided with a trap in the escape-pipe for the recovery of small objects which may have been accidentally dropped down. Behind the table is seen a strong wooden partition, open at the top, separating it from similar tables at the other side. In front of this are fixed, as shown, tall iron upright rods of the same thickness as the rods of the retort stands, and to which the retort stand clips can be fixed, at any desired height, for holding pendulums, torsion wires, etc. For other tables, these uprights are fixed to the walls of the room.

It is convenient to have the front cupboards fitted with snap-locks, all of which can be opened by the same key. In order to secure complete tidiness in the cupboards, I have found it essential that every piece of apparatus should have its place duly marked with a label in the cupboard, and to insist that each piece shall be put back exactly in its right place. Three or four minutes before the end of the lesson the word is given to put away the apparatus, and either before or after the boys have gone, the master, walking round and looking at the open cupboards, can tell at a glance whether everything is in its place. The cupboard doors are snapped to, and five minutes after the cessation of work he can walk out of the room, assured that everything is in perfect order for the next lesson. The apparatus not in immediate use is kept in suitable cupboards on the walls of the room. A complete list of all that is required for the experiments of this course will be found at the end of the book.

A few words on the question of cost of this kind of laboratory work may not be out of place.

The tables that I have described cost £3, 3s. each, and the total cost of the interior fittings of the Clifton laboratory, for a class of thirty boys, including the tables, wall-cupboards, blackboard, a stone sink, central partition, gas and water fittings, may be put at very nearly £70. The original outlay in apparatus four years ago was £40, and the annual cost of working, supplying material, and replacing breakages, has for the last three years been £12, 10s. per annum.

With respect to the time required for the performance of the experiments described in the course, the following details may be useful; the information given refers to pupils of ordinary intelligence and industry.

Part I. (Mensuration) requires a little more than one school term of thirteen lessons of one hour each, though it is sometimes finished in one term.
Parts I., II., and III. (Mensuration, Hydrostatics, and Barometer and Boyle's Law) can be accomplished in two such terms.
Part IV. (Mechanics, including (1) Levers and Parallelogram of Forces, (2) Centre of Gravity, (3) Friction and Inclined Plane, (4) Friction of Cords and Use of Pulleys) takes less than one term of twenty-six hour lessons.
Part V. (Elasticity and the law of the Simple Pendulum) takes less than one term of twenty-six hours.
Part VI. (Heat) requires about one term of twenty-six hours.
Part VII. (Magnetism) takes rather less than one term of twenty-six hours.
Part VIII. (Statical Electricity) takes one term of twenty-six hours.
Part IX. (Current Electricity) takes about one term of twenty-six hours.

The order in which the sections are taken by the pupil need not necessarily be that in which they are arranged in the book, but it is recommended that the first three sections be imperative on all beginners, and that they be taken in order.

An acquaintance with vulgar and decimal fractions and simple proportion, in Arithmetic, and with at least the first book of Euclid, is requisite.

The following Rules to be observed by Boys working in the Laboratory have been found convenient.

RULES

1. Read all about the experiment, so as to have a clear idea of what you are going to do before you begin.
2. Remove all apparatus from the table to its proper place *as soon as it is done with.*
3. The form in which the experiment is to be recorded should be got ready beforehand, so that as soon as an observation is made it can be recorded in its right place in the form.
4. The numbers thus recorded in pencil in the rough note-book must not be the result of additions or subtractions done in the head, but the numbers on which such calculations, however short, depend.
5. Even when the measure made by the second observer of a pair agrees with that made by the first, both must be recorded.
6. Make all the necessary observations before proceeding to any calculation.
7. Books, etc., not wanted during the lesson must not be placed on the tables.

8. All waste-paper, etc., to be put in the waste-baskets.
9. All breakages to be reported at once, and put down in the breakage-book.
10. No talking is allowed, except in a low voice to your partner.
11. Before leaving the room, examine your cupboard to see that everything is in its place.

Editorial Headnote

Horace Lemuel Wells, *A Laboratory Guide in Qualitative Chemical Analysis* (New York: Wiley, 1898), iii–iv

Horace Lemuel Wells, 1855–1924, was professor of analytical chemistry at Yale. The introduction to his texts for students, itself the product of his own experience in the laboratory, emphasises the need for young experimenters to re-discover, through their own experiments, already known facts for themselves.

17

A LABORATORY GUIDE IN QUALITATIVE CHEMICAL ANALYSIS

Horace Lemuel Wells

Source: New York: Wiley, 1898, iii–iv

THE method of instruction presented in this little book has been developed by the author in teaching the subject during the past fourteen years. The course differs materially from those usually employed, and has given satisfaction because of its tendency to obviate thoughtless and mechanical work on the part of the student—facts which may be offered as an excuse for adding another to the great number of works on qualitative analysis. The general plan of Part I, which is applied wherever it would not be too cumbersome, consists in giving directions for analyses without mentioning the results, and in requiring the student to determine for himself the results of the operations. Following this system, the student almost invariably performs his preliminary experiments with great care, because he recognizes their object and importance. This is especially the case if the student understands that, in the analysis of unknown mixtures, he will not be allowed to use any tables or abbreviated schemes of analysis.

The methods of analysis given by Fresenius are chiefly used, but a number of changes have been made where experience has shown them to be advantageous on account of simplicity or accuracy. A few novelties are introduced in the way of details of processes or combinations of methods, but it may be stated that these have been thoroughly tested and may be relied upon as trustworthy. An original method, now published for the first time, is used for the detection of chlorides in the presence of bromides. Lithium has been included among the commoner metals, on account of the extensive use of its salts in medicine and their commercial abundance. This addition considerably increases the interest of the alkali-metal group.

The book is not intended to be a complete and independent work on analytical chemistry, and students who use it should be encouraged to increase their knowledge of the subject by frequent reference to a comprehensive text-book.

Before beginning the study of qualitative analysis, the student should possess a good knowledge of elementary chemistry, including the fundamental chemical

DOI: 10.4324/9781003009337-23

theories; but there is much theoretical knowledge to be gained, or more firmly grasped, in connection with the study of analysis. On this account, a section of the book (Part II) is devoted to the explanation of the formation of formulas and equations, and to some other points of theory which have a special bearing upon the course of study, including the important subject of ionization.

Part III contains an alphabetical arrangement of the radicals, with brief descriptions of the properties upon which the analytical methods are based. Most of the facts given there should be familiar to the student after completing the course presented in Part I, so that it will be useful in reviewing the subject. This part may be of use, also, for reference in regard to reactions, the formulas of precipitates, etc., and as an index to the fuller descriptions in Fresenius's Manual.

The author is under obligation to Dr. B. B. Boltwood and Mr. H. W. Foote, Instructors in the Sheffield Laboratory, for many valuable suggestions, and to Miss L. P. Bush for assistance in proof-reading.

Editorial Headnote

Marie Skłodowska-Curie, 'Radium and Radioactivity', *Century Magazine* (1904), 461–466

Marie Skłodowska-Curie, 1867–1934, is one of the most well-known figures in the history of science, being the first to win a Nobel Prize twice, for the development of the theory of radioactivity and for the discovery of polonium and radium, respectively. She was also the first woman to win the prize. The text here reflects on the limits of what can be seen or perceived, and the need to contrive methods to make knowledge out of the invisible.

18

'RADIUM AND RADIOACTIVITY'

Marie Skłodowska-Curie

Source: *Century Magazine* (1904), 461–466

THE discovery of the phenomena of radioactivity adds a new group to the great number of invisible radiations now known, and once more we are forced to recognize how limited is our direct perception of the world which surrounds us, and how numerous and varied may be the phenomena which we pass without a suspicion of their existence until the day when a fortunate hazard reveals them.

The radiations longest known to us are those capable of acting directly upon our senses; such are the rays of sound and light. But it has also long been recognized that, besides light itself, warm bodies emit rays in every respect analogous to luminous rays, though they do not possess the power of directly impressing our retina. Among such radiations, some, the infra-red, announce themselves to us by producing a measurable rise of temperature in the bodies which receive them, while others, the ultra-violet, act with specially great intensity upon photographic plates. We have here a first example of rays only indirectly accessible to us.

Yet further surprises in this domain of invisible radiations were reserved for us. The researches of two great physicists, Maxwell and Herz, showed that electric and magnetic effects are propagated in the same manner as light, and that there exist "electromagnetic radiations," similar to luminous radiations, which are to the infra-red rays what these latter are to light. These are the electromagnetic radiations which are used for the transmission of messages in wireless telegraphy. They are present in the space around us whenever an electric phenomenon is produced, especially a lightning discharge. Their presence may be established by the use of special apparatus, and here again the testimony of our senses appears only in an indirect manner. If we consider these radiations in their entirety,—the ultra-violet, the luminous, the infra-red, and the electromagnetic,—we find that the radiations we see constitute but an insignificant fraction of those that exist in space. But it is human nature to believe that the phenomena we know are the only ones that exist, and whenever some chance discovery extends the limits of our knowledge we are filled with amazement. We cannot become accustomed to the idea that we live in a world that is revealed to us only in a restricted portion of its manifestations.

DOI: 10.4324/9781003009337-24

Among recent scientific achievements which have attracted most attention must be placed the discovery of cathode rays, and in even greater measure that of Roentgen rays. These rays are produced in vacuum-tubes when an electric discharge is passed through the rarefied gas. The prevalent opinion among physicists is that cathode rays are formed by extremely small material particles, charged with negative electricity, and thrown off with great velocity from the cathode, or negative electrode, of the tube. When the cathode rays meet the glass wall of the tube they render it vividly fluorescent. These rays can be deflected from their straight path by the action of a magnet. Whenever they encounter a solid obstacle, the emission of Roentgen rays is the result. These latter can traverse the glass and propagate themselves through the outside air. They differ from cathode rays in that they carry no electric charge and are not deflected from their course by the action of a magnet. Every one knows the effect of Roentgen rays upon photographic plates and upon fluorescent screens, the radiographs obtainable from them, and their application in surgery.

The discovery of Becquerel rays dates from a few years after that of Roentgen rays. At first they were much less noticed. The world, attracted by the sensational discovery of Roentgen rays, was less inclined to astonishment. On all sides a search was instituted by similar processes for new rays, and announcements of phenomena were made that have not always been confirmed. It has been only gradually that the positive existence of a new radiation has been established. The merit of this discovery belongs to M. Becquerel, who succeeded in demonstrating that uranium and its compounds spontaneously emit rays that are able to traverse opaque bodies and to affect photographic plates.

It was at the close of the year 1897 that I began to study the compounds of uranium, the properties of which had greatly attracted my interest. Here was a substance emitting spontaneously and continuously radiations similar to Roentgen rays, whereas ordinarily Roentgen rays can be produced only in a vacuum-tube with the expenditure of electrical energy. By what process can uranium furnish the same rays without expenditure of energy and without undergoing apparent modification? Is uranium the only body whose compounds emit similar rays? Such were the questions I asked myself, and it was while seeking to answer them that I entered into the researches which have led to the discovery of radium.

First of all, I studied the radiation of the compounds of uranium. Instead of making these bodies act upon photographic plates, I preferred to determine the intensity of their radiation by measuring the conductivity of the air exposed to the action of the rays. To make this measurement, one can determine the speed with which the rays discharge an electroscope, and thus obtain data for a comparison. I found in this way that the radiation of uranium is very constant, varying neither with the temperature nor with the illumination. I likewise observed that all the compounds of uranium are active, and that they are more active the greater the proportion of this metal which they contain. Thus I reached the conviction that the emission of rays by the compounds of uranium is a property of the metal itself—that it is an *atomic property of the element uranium* independent of its chemical or physical state.

I then began to investigate the different known chemical elements, to determine whether there exist others, besides uranium, that are endowed with atomic radioactivity—that is to say, all of the compounds of which emit Becquerel rays. It was easy for me to procure samples of all the ordinary substances—the common metals and metalloids, oxids and salts. But as I desired to make a very thorough investigation, I had recourse to different chemists, who put at my disposal specimens—in some cases the only ones in existence—containing very rare elements. I thus was enabled to pass in review all the chemical elements and to examine them in the state of one or more of their compounds. I found but one element undoubtedly possessing atomic radioactivity in measurable degree. This element is thorium. All the compounds of thorium are radioactive, and with about the same intensity as the similar compounds of uranium. As to the other substances, they showed no appreciable radioactivity under the conditions of the test.

I likewise examined certain minerals. I found, as I expected, that the minerals of uranium and thorium are radioactive; but to my great astonishment I discovered that some are much more active than the oxids of uranium and of thorium which they contain. Thus a specimen of pitch-blende (oxid of uranium ore) was found to be four times more active than oxid of uranium itself. This observation astonished me greatly. What explanation could there be for it? How could an ore, containing many substances which I had proved inactive, be more active than the active substances of which it was formed? The answer came to me immediately: The ore must contain a substance more radioactive than uranium and thorium, and this substance must necessarily be a chemical element as yet unknown; moreover, it can exist in the pitch-blende only in small quantities, else it would not have escaped the many analyses of this ore; but, on the other hand, it must possess intense radioactivity, since, although present in small amount, it produces such remarkable effects. I tried to verify my hypothesis by treating pitch-blende by the ordinary processes of chemical analysis, thinking it probable that the new substance would be concentrated in passing through certain stages of the process. I performed several experiments of this nature, and found that the ore could in fact be separated into portions some of which were much more radioactive than others.

To try to isolate the supposed new element was a great temptation. I did not know whether this undertaking would be difficult. Of the new element I knew nothing, except that it was radioactive. What were its chemical properties? In what quantity did it appear in pitch-blende? I began the analysis of pitch-blende by separating it into its constituent elements, which are very numerous. This task I undertook in conjunction with M. Curie. We expected that perhaps a few weeks would suffice to solve the problem. We did not suspect that we had begun a work which was to occupy years and which was brought to a successful issue only after considerable expenditure.

We readily proved that pitch-blende contains very radioactive substances, and that there were at least three. That which accompanies the bismuth extracted from pitch-blende we named Polonium; that which accompanies barium from the same source we named Radium; finally, M. Debierne gave the name of Actinium to a substance which is found in the rare earths obtained from the same ore.

Radium was to us from the beginning of our work a source of much satisfaction. Demarçay, who examined the spectrum of our radioactive barium, found in it new rays and confirmed us in our belief that we had indeed discovered a new element.

The question now was to separate the polonium from the bismuth, the radium from the barium. This is the task that has occupied us for years, and as yet we have succeeded only in the case of radium. The research has been a most difficult one. We found that by crystallizing out the chlorid of radioactive barium from a solution we obtained crystals that were more radioactive, and consequently richer in radium, than the chlorid that remained dissolved. It was only necessary to make repeated crystallizations to obtain finally a pure chlorid of radium.

But although we treated as much as fifty kilograms of primary substance, and crystallized the chlorid of radiferous barium thus obtained until the activity was concentrated in a few minute crystals, these crystals still contained chiefly chlorid of barium; as yet radium was present only in traces, and we saw that we could not finish our experiments with the means at hand in our laboratory. At the same time the desire to succeed grew stronger; for it became evident that radium must possess most intense radioactivity, and that the isolation of this body was therefore an object of the highest interest.

Fortunately for us, the curious properties of these radium-bearing compounds had already attracted general attention, and we were assisted in our search.

A chemical factory in Paris consented to undertake the extraction of radium on a large scale. We also received certain pecuniary assistance, which allowed us to treat a large quantity of ore. The most important of these grants was one of twenty thousand francs, for which we are indebted to the Institute of France.

We were thus enabled to treat successively about seven tons of a primary substance which was the residue of pitch-blende after the extraction of uranium. To-day we know that a ton of this residue contains from two to three decigrams (from four to seven ten-thousandths of a pound) of radium. During this treatment, and as soon as I had in my possession a decigram of chlorid of radium recognized as pure by the spectroscope, I determined the atomic weight of this new element, finding it to be 225, while that of barium is 137.

The properties of radium are extremely curious. This body emits with great intensity all of the different rays that are produced in a vacuum-tube. The radiation, measured by means of an electroscope, is at least a million times more powerful than that from an equal quantity of uranium. A charged electroscope placed at a distance of several meters can be discharged by a few centigrams of a radium salt. One can also discharge an electroscope through a screen of glass or lead five or six centimeters thick. Photographic plates placed in the vicinity of radium are almost instantly affected if no screen intercepts the rays; with screens, the action is slower, but it still takes place through very thick ones if the exposure is sufficiently long. Radium can therefore be used in the production of radiographs.

The compounds of radium are *spontaneously luminous*. The chlorid and bromide, freshly prepared and free from water, emit a light which resembles that of the glowworm. This light diminishes rapidly in moist air; if the salt is in a sealed tube, it diminishes slowly by reason of the transformation of the white salt, which becomes colored, but the light never completely disappears. By redissolving the salt and drying it anew, its original luminosity is restored.

A glass vessel containing radium *spontaneously charges itself with electricity*. If the glass has a weak spot,—for example, if it is scratched by a file,—an electric spark is produced at that point, the vessel crumbles like a Leiden jar when overcharged, and the electric shock of the rupture is felt by the fingers holding the glass.

Radium possesses the remarkable property *of liberating heat spontaneously and continuously*. A solid salt of radium develops a quantity of heat such that for each gram of radium contained in the salt there is an emission of one hundred calories per hour. Expressed differently, radium can melt in an hour its weight in ice. When we reflect that radium acts in this manner *continuously*, we are amazed at the amount of heat produced, for it can be explained by no known chemical reaction. The radium remains apparently unchanged. If, then, we assume that it undergoes a transformation, we must therefore conclude that the change is extremely slow: in an hour it is impossible to detect a change by any known methods.

As a result of its emission of heat, radium always possesses a higher temperature than its surroundings. This fact may be established by means of a thermometer, if care is taken to prevent the radium from losing heat.

Radium has the power of communicating its radioactivity to surrounding bodies. This is a property possessed by solutions of radium salts even more than by the solid salts. When a solution of a radium salt is placed in a closed vessel, the radioactivity in part leaves the solution and distributes itself through the vessel, the walls of which become radioactive and luminous. The radiation is therefore in part exteriorized. We may assume, with Mr. Rutherford, that radium emits a radioactive gas and that this spreads through the surrounding air and over the surface of neighboring objects. This gas has received the name *emanation*. It differs from an ordinary gas in the fact that it gradually disappears.

Certain bodies—bismuth, for instance—may also be rendered active by keeping them in solution with the salts of radium. These bodies then become atomically active and keep this radioactivity even after chemical transformations. Little by little, however, they lose it, while the activity of radium persists.

The nature of radium radiations is very complex. They may be divided into three distinct groups, according to their properties.

One group is composed of radiations absolutely analogous to cathode rays, composed of material particles called electrons, much smaller than atoms, negatively charged, and projected from the radium with great velocity—a velocity which for some of these rays is very little inferior to that of light.

The second group is composed of radiations which are believed to be formed by material particles the mass of which is comparable to that of atoms, charged with positive electricity, and set in motion by radium with a great velocity, but one that is inferior to that of the electrons. Being larger than electrons and possessing at the same time a smaller velocity, these particles have more difficulty in traversing obstacles and form rays that are less penetrating.

Finally, the radiations of the third group are analogous to Roentgen rays and do not behave like projectiles.

The radiations of the first group are easily deflected by a magnet; those of the second group, less easily and in the opposite direction; those of the third group are not deflected. From its power of emitting these three kinds of rays, radium may be likened to a complete little Crookes tube acting spontaneously.

Radium is a body which gives out energy continuously and spontaneously. This liberation of energy is manifested in the different effects of its radiation and emanation, and especially in the development of heat. Now, according to the most fundamental principles of modern science, the universe contains a certain definite provision of energy, which can appear under various forms, but cannot be increased. Without renouncing this conception, we cannot believe that radium creates the energy which it emits; but it can either absorb energy continuously from without, or possess in itself a reserve of energy sufficient to act during a period of years without visible modification. The first theory we may develop by supposing that space is traversed by radiations that are as yet unknown to us, and that radium is able to absorb these radiations and transform their energy into the energy of radioactivity. Thus in a vacuum-tube the electric energy is utilized to produce cathode rays, and the energy of the latter is partly transformed, by the bodies which absorb them, into the energy of Roentgen rays. It is true that we have no proof of the existence of radiations which produce radioactivity; but, as indicated at the beginning of this article, there is nothing improbable in supposing that such radiations exist about us without our suspecting it.

If we assume that radium contains a supply of energy which it gives out little by little, we are led to believe that this body does not remain unchanged, as it appears to, but that it undergoes an extremely slow change. Several reasons speak in favor of this view. First, the emission of heat, which makes it seem probable that a chemical reaction is taking place in the radium. But this can be no ordinary chemical reaction, affecting the combination of atoms in the molecule. No chemical reaction can explain the emission of heat due to radium. Furthermore, radioactivity is a property of the atom of radium; if, then, it is due to a transformation, this transformation must take place in the atom itself. Consequently, from this point of view, the atom of radium would be in a process of evolution, and we should be forced to abandon the theory of the invariability of atoms, which is at the foundation of modern chemistry.

Moreover, we have seen that radium acts as though it shot out into space a shower of projectiles, some of which have the dimensions of atoms, while others can only be very small fractions of atoms. If this image corresponds to a reality, it follows necessarily that the atom of radium breaks up into subatoms of different sizes, unless these projectiles come from the atoms of the surrounding gas, disintegrated by the action of radium; but this view would likewise lead us to believe that the stability of atoms is not absolute.

Radium emits continuously a radioactive emanation which, from many points of view, possesses the properties of a gas. Mr. Rutherford considers the emanation as one of the results of the disintegration of the atom of radium, and believes it to be an unstable gas which is itself slowly decomposed.

Professor Ramsay has announced that radium emits helium gas continuously. If this very important fact is confirmed, it will show that a transformation is occurring either in the atom of radium or in the neighboring atoms, and a proof will exist that the transmutation of the elements is possible.

When a body that has remained in solution with radium becomes radioactive, the chemical properties of this body are modified, and here again it seems as though we have to deal with a modification of the atom. It would be very interesting to see whether, by thus giving radioactivity to bodies, we can succeed in causing an appreciable change in their atoms. We should thus have a means of producing certain transformations of elements at will.

It is seen that the study of the properties of radium is of great interest. This is true also of the other strongly radioactive substances, polonium and actinium, which are less known because their preparation is still more difficult. All are found in the ores of uranium and thorium, and this fact is certainly not the result of chance, but must have some connection with the manner of formation of these elements. Polonium, when it has just been extracted from pitch-blende, is as active as radium, but its radioactivity slowly disappears; actinium has a persistent activity. These two bodies differ from radium in many ways; their study should therefore be fertile in new results. Actinium lends itself readily to the study of the emanation and of the radioactivity produced in inactive bodies, since it gives out emanation in great quantity. It would also be interesting, from the chemical point of view, to prove that polonium and actinium contain new elements. Finally, one might seek out still other strongly radioactive substances and study them.

But all these investigations are exceedingly difficult because of the obstacles encountered in the preparation of strongly radioactive substances. At the present time we possess only about a gram of the pure salts of radium. However great may be the care taken in such researches, small losses are inevitable, and serious losses have at times resulted from unforeseen accidents brought on by the disconcerting properties of radium. Research in all branches of experimental science—physics, chemistry, physiology, medicine—is impeded, and a whole evolution in science is retarded, by the lack of this precious and unique material, which can now be obtained only at great expense. We must now look to individual initiative to come to the aid of science, as it has so often done in the past, and to facilitate and expedite by generous gifts the success of researches the influence of which may be far-reaching.

Editorial Headnote

Ernst von Meyer, *A History of Chemistry From Earliest Times to the Present Day* (London: MacMillan, 1906), 6, 96–9, 217–18

Ernst von Meyer, 1847–1916, was a German professor of chemistry. This text reflects on the long history of chemistry and the shifting fortunes of the status of experimentation since ancient times, with corresponding effects on the status of knowledge.

19

A HISTORY OF CHEMISTRY FROM EARLIEST TIMES TO THE PRESENT DAY

Ernst von Meyer

Source: London: MacMillan, 1906, 6, 96–99, 217–218

In sharp contrast with the disinclination of the Ancients towards experiment, by which the secrets of nature are to be unravelled, stands their great love of speculation, by means of which they did not hesitate to attempt an explanation of the ultimate reasons of all things. Aristotle, to whom the natural sciences owed the direction which they followed for a long time, pointed to deduction as the road which should lead to the goal. Instead of drawing general conclusions from accurately observed facts, the Ancients preferred to advance from the general to the particular. The position of all the natural sciences in farback times, especially that of chemistry, is sufficient to prove how the most mischievous errors crept in and became firmly established in consequence of the purely deductive method.

The philosophical writings of the Ancients, especially those of the Greeks and Romans, give us a tolerably distinct idea of their theoretical views. Certain writings of Aristotle, and also the "*περὶ λίθων*" of his pupil Theophrastus, are of especial value for the criticism of the empirical chemical knowledge of these times. . . .

HISTORY OF THE PERIOD OF THE PHLOGISTON THEORY, FROM BOYLE TO LAVOISIER

. . . THE reasons for naming this period of about one hundred and twenty years the period of the phlogiston theory, or of phlogistic chemistry, have been already stated shortly (p. 4). For the first portion of this era the designation is in truth not absolutely fitting, since Robert Boyle—the man who above all others gave a new direction to chemistry at the time—did not concur in the phlogistic views. The development proper of the phlogiston theory really took place after his death. Nevertheless the period from Boyle to Lavoisier may be so named, because the most

DOI: 10.4324/9781003009337-25

important part of chemical research during that time had to do with the phenomena of combustion and with the analogous calcination of the metals. All the most eminent chemists of that day directed their attention to this problem, both theoretically and experimentally. It formed, especially towards the end of this period, the centre around which the whole of chemistry circled; it became a stumbling-block to the adherents of the old doctrines, and led to a reform of the science so fundamental and far-reaching that the chemistry of to-day still lives under it.

The iatro-chemical theories strove after the impossible, and therefore quickly succumbed; the marked one-sidedness apparent in them, the gratuitous explanations of life-processes, and the total neglect of the anatomy and morphology of the organs, made their decline inevitable. An opportunity was thus given to chemistry to loosen and finally break the bands which medicine had wound around her, and to take up an independent position of her own. She still remained for a time under the protection of the healing art, to which she was indeed an indispensable aid; but even from the time of Boyle on, the aim of chemistry was recognised as being the discovery of new chemical facts, purely for the sake of arriving at the truth alone.

The spirit of true investigation which penetrated the natural sciences at the end of the sixteenth and beginning of the seventeenth centuries, also began to impart itself to chemistry, the development of physics exerting an especially powerful influence upon the younger sister-science. The inductive method also acquired, as a guide, a continually growing and a lasting influence, the nature of which was indicated by Francis Bacon[1] substantially as follows:—

> "The true kind of experience is not the mere groping of a man in the dark, who feels at random to find his way, instead of waiting for the dawn or striking a light . . . . It begins with an ordered—not chaotic—knowledge of facts, deduces axioms from these, and from the axioms again designs new experiments." Equipped with such axioms, chemistry might enrol itself among the exact sciences.

The learned societies which came into existence in the second half of the seventeenth and beginning of the eighteenth centuries, and whose periodicals spread abroad the results of chemical investigations, aided materially towards the healthy development of the science. The incitement given by them to researches, which could then be submitted to verification by other workers, was also not to be undervalued. Finally, they promoted the reciprocal action of chemistry and allied branches of science upon each other, an action so fruitful in its results, by bringing their exponents into closer connection.

The Royal Society, which was formed about the middle of the seventeenth century by the amalgamation of the two smaller scientific societies of Oxford and London, and which began to publish the *Philosophical Transactions* in 1665, furnishes a good instance of what has just been said. The Italian academies, especially the *Academia del Cimento* of Florence (1657), devoted themselves mostly to physical and mathematical studies. In Vienna the *Academia Naturæ Curiosorum* was started in 1652, taking the name of *Cæsarea Leopoldina* in honour of its patron Leopold I. The *Académie Royale* originated in Paris in 1666 out of friendly meetings which were held at the house of the physicist Mersenne; the *Mémoires de l'Académie des Sciences* began to appear in 1699. The Berlin Academy was founded in 1700 by Frederick I., Leibniz being its first president; and during the first half of the eighteenth century the northern countries followed suit with similar learned societies, that of St. Petersburg being started in 1725, that of Stockholm in 1739, and that of Copenhagen in 1743.

That an extraordinary interest was felt at this time in scientific questions is readily seen from the literature of the day, which reflects the excitement—sometimes feverish in its intensity—raised by isolated discoveries, like that of phosphorus, or by disputed problems, such as the question of the cause of combustion.

The modes in which chemical questions were treated did indeed approximate to the methods followed in recent times, but in one respect there was a wide distinction between them. The chemical investigation of the phlogistic period took very little note of the proportions by weight in which substances entered into reaction; it turned its attention almost alone to the qualitative side of the phenomena. The setting up and the subsequent development of the phlogistic doctrines were only possible because of the utter neglect of quantitative relations. Even sagacious thinkers who, by observing that metals increased in weight upon calcination, came thus into direct conflict with the phlogistic view, evaded the only correct explanation of this,—and, with it, of the phenomena of combustion—by far-fetched conceptions. This blinding of the mind by an erroneous theory, consequent upon the refusal to look into all the conditions which might have contributed to clearing up the question, is peculiar to the period of phlogistic chemistry.

In spite, however, of the fundamental error which ran through it, this period was a highly fruitful one for chemistry; it forms the indispensable introduction to the most recent phase of development of the science. And although it was itself fettered by many erroneous ideas, still the phlogistic age contributed very largely to the refutation of mischievous errors, *e.g.* those belonging to the iatro-chemical doctrines and the false beliefs of alchemy. . . .

. . . The experimental researches of Berzelius were as a rule first published in Swedish in the *Transactions of the Stockholm Academy*, but most of them were afterwards given out in German, and a few in English and French (in *Gilbert's*, *Poggendorff's*, and *Liebig's Annalen*, the *Annales de Chimie*, *Annals of Philosophy*, etc.). They are characterised by the same excellences as his text-book.

The above sketch of his main achievements is sufficient to indicate the eminent qualities which distinguished Berzelius as a classical investigator. Thoroughness and perseverance in everything which he undertook; exactness in all his observations, and the capacity for arranging these distinctly and explaining them clearly; inviolable adherence to the results of experience (which was his guide before everything else), and an equally firm adherence to results which, in his opinion, had been correctly arrived at from a number of data; these were the characteristics which distinguished this great man.

The desire to retain whatever of good the science possessed was developed in him in an exceptional degree; indeed, in sustaining this conservative attitude he went so far as to see a danger to the steady development of chemistry in every innovation which called in question views already proved and found useful. Hence his fervent opposition to many new hypotheses which he had in the end to recognise as correct. His great services in furthering chemistry were, however, not lessened by this peculiarity, which had its real cause in a profound sense of justice; on the contrary, by a prudent adherence to approved opinions, Berzelius often prevented the confusion to which the views he combated might have given rise, had they been accepted without reservation. Not that he was averse to healthy reform. But against anything violent—to his mind revolutionary—he fought with all his energy; he did not shun even hot polemics[2] when anything that he regarded as sound was at stake.

His pupil Heinrich Rose gave a comprehensive review of his general character in the "Memorial Speech of Berzelius,"[3]—a speech of great beauty and with a pleasant warmth of tone running through it. At the close of it (p. 59) Rose says: "The irresistible captivation which Berzelius exercised over those who enjoyed the privilege of a lengthened intercourse with him was only partly due to the lofty genius, whose sparks flashed from all his work, and only partly to the clearness, the marvellous wealth of ideas, and the untiring care and great industry that gave everything with which he had to do the stamp of the highest perfection. It was also—and every one who knew him intimately will agree with me in this,—it was also those qualities which placed him so high as a man; it was his devotion to others, the noble friendship which he showed to all whom he deemed worthy of it, the great unselfishness and conscientiousness, the perfect and just recognition of the services of others,—in short, it was all those qualities which spring from an upright and honourable character."

Notes

1 *Novum Organon,* Aphorism 82, paragraph 3. Bacon in the above paragraph gave expression to no new idea, but merely called special attention to the value of experience, a point already recognised by his predecessors Palissy, Leonardo da Vinci, Paracelsus, and others. Liebig in a series of essays has proved conclusively how unjustifiable it is to designate Bacon as the originator of the inductive method, and how little he was permeated by the spirit of true research (see his *Reden und Abhandlungen,* 1874).

2 His controversies with Dumas, Laurent, Liebig, and others have often been harshly and unfairly criticised, in that a false light has been thereby thrown upon his whole work. The younger generation of chemists, in especial, quickly forgot after his death the debt which was due to him for the imperishable services which he had rendered in the building up of the science. Indeed, derision and cheap ridicule of the mistakes he made are still to be found in recent works which treat of the development of chemical theories.

3 Delivered at a public meeting of the Berlin Academy, 3d July 1851.

Editorial Headnote

Svante Arrhenius, *Theories of Chemistry: Being Lectures Delivered at the University of California, in Berkeley* (London: Longmans, Green, and Co., 1907), 1–11

Svante August Arrhenius, 1859–1927, was a Swedish physicist and chemist, and recipient of the Nobel Prize for Chemistry in 1903, for his work on conductivities of electrolytes. Here he presents the relation of theory to experimentation.

20

THEORIES OF CHEMISTRY

Being Lectures Delivered at the University of California, in Berkeley

Svante Arrhenius

Source: London: Longmans, Green, and Co., 1907, 1–11

THEORIES OF CHEMISTRY

Chapter I

Introduction: the use of theories

In the following lectures I shall have the honour of bringing before you some account of the development of theories in chemistry.

To commence, then, at the beginning, we have to ask ourselves: "What is the characteristic feature of a theory?" Most people would answer: "A theory is something unpractical." Nothing could be more incorrect. It is just the contrary. As Boltzmann once said, "A theory should be the most practical exposition of facts." As an example he cites the theory of the conduction of electricity, the greater part of which is embodied in Ohm's law. Before Ohm had formulated his law, which he deduced from theoretical considerations, an endless number of experiments were made to ascertain the connection between the strength of a current (estimated either according to the brightness of the sparks produced on breaking the circuit, or from the deflection of a magnet) and the number of voltaic cells used for its production. The cells, however, did not possess a constant resistance, and their electromotive force changed rapidly with the time, so that it was very difficult to deduce the law, expressing the connection sought for, from purely experimental data.

Ohm,[1] however, applied deductions which he derived mainly from the results of Fourier's classical researches on the conduction of heat. Just as the quantity of heat flowing through a parallelopiped, the ends of which are maintained at the constant temperatures t_1 and t_0 respectively, is proportional to the cross-section and inversely proportional to the length of the parallelopiped, also proportional to the difference of temperature $(t_1 - t_0)$ and to a constant, called the specific

DOI: 10.4324/9781003009337-26

conductivity for heat, characteristic of the substance of which the parallelopiped is made; so, in the same way, the current of electricity through a parallelopiped (*e.g.* a cylinder or a wire) ought to be proportional to the cross-section and inversely proportional to the length of the parallelopiped, also proportional to the difference of electrical potential between its end surfaces and to a constant, called the specific electrical conductivity, characteristic of the substance. In this way Ohm arrived at his law, and all experiments which have since been made on the strength of electric currents have verified it.

Now, even if Ohm's law had only been true for weak currents, and deviations from it had been found for strong currents, it would nevertheless have been extremely useful. Because of its *quantitative* formulation, it was possible to calculate the strength of a given current and compare this calculated value with the observed one. If now both had not agreed within the limits of experimental error, it would have been necessary to seek for the cause of this deviation. In this way we should have found either that the electromotive force (*i.e.* the potential difference) applied to the ends of the wire was not that used in the calculation, or that a wrong value had been assumed for the constant representing the electrical conductivity of the wire. We should then have made a new experiment, taking care that the electromotive force and the conductivity were of the supposed values, and then observed if the value of the current agreed with that of the previous observation or with the calculated value, within the limits of experimental error. In the latter case an experimental proof of the truth of the law would have been given. This proof has been obtained by so many observers that if the former case were ever to occur we should conclude that some hitherto unknown phenomenon had been met with, or that some error had been made in the measurements. In order to decide between these alternatives it would be necessary to make new [Illegible Text] still more careful experiments. Without the theory there then should have been no reason for new and more accurate research. [Illegible Text] general, after a theory (expressed concisely by means of a formula) has been verified in a great number of cases, it is of the highest practical value, for it permits us to calculate—much more accurately than we could determine experimentally—the magnitude of one of the quantities (*e.g.* the strength of current) represented in the formula, when the others are known. Thus theory is the greatest promoter of scientific research.

Returning to our consideration of Ohm's law, we find that it leads to researches on the influence of temperature, pressure, and time on the conductivity of different substances, and on the effect of dilution on the conductivity of electrolytes. In recent times such researches have resulted in the development of the theory of electrolytic dissociation. In this case Ohm's law was used, together with other laws deduced from van't Hoff's theory of the freezing-point of solutions.

Before a theory is established, we generally find an hypothesis used, which later on may develop into a theory. Such an hypothesis was, for instance, the assumption that it is the electromotive force (proportional to the number of cells used) that drives the electric current, which may be regarded as the motion of a substance through the conducting wires. (At the time of Ohm's discovery heat and electricity were regarded as matter.) All researches previous to Ohm's on the connection between the number of cells acting and the strength of the current must have been based on this hypothesis.

The work of Ohm transformed this rather vague and incomplete hypothesis into a law. *Quantitative* formulation, that is, the establishing of a connection, expressed by a formula, between different quantitatively measurable magnitudes, is the peculiar feature of a law or a theory. Of course, this is only valid for the so-called exact sciences, which deal with measurable *quantities*. A purely descriptive science develops to an exact one in so far as it introduces theories in the above sense, and it is just the development of chemistry in this direction in the last century which may be regarded as one of the best illustrations of our statement.

The introduction of statistical methods in biological science as has been done in modern physiological researches on the development of living organisms, permits of the establishment theories in these branches of knowledge. In biological sciences the word "theory" is often used in the same sense as hypothesis still, this has no influence upon our study of theories in exact sciences.

The main feature of a theory is, as we have just said, its bearing upon "quantities." It follows that measuring instruments must have been introduced before a theory can be established. In the case of Ohm's theory of the electric current, such an instrument was the galvanometer which had just been constructed by Schweigger.

According to the definition given above, the so-called phlogiston theory does not at all merit its name, for no instrument was devised for measuring the quantity of phlogiston, by the combination of which with the calces (oxides) metals were supposed to be produced. It was only an hypothesis which might have been developed into a theory, if some one had measured phlogiston, after working out a clear conception of its nature. Thus, for instance, it would have been possible to put the quantity of phlogiston given off on the oxidation of a metal equal to the quantity of heat developed during this process, the heat being measured calorimetrically. (In reality the old conception of phlogiston seems most nearly to correspond to this quantity of heat.) But instead of using a calorimeter, Lavoisier employed the balance for investigating the phenomenon of calcination; and he found that just as much oxygen disappears during the calcination of a metal (he used tin, lead, and quicksilver) as the metal gains in weight.

He therefore propounded the theory that calcination consists of the combination of a metal with oxygen, and he demonstrated his theory by quantitative measurements. With this theory the old phlogiston hypothesis, lacking all quantitative foundation, did not agree, and was therefore given up.

The quantitative use of the balance resulted afterwards in the atomic theory of Dalton, of which the most important feature is the law of multiple proportions. This theory was a development of the atomic hypothesis of Democritus, which had already been known to science for about 2300 years.

[Illegible Text] on the [Illegible Text] have already seen, theories have a practical side of [Illegible Text] of importance in that they economize experimental work. We may compare the work of an investigator, who wishes to find the connection between two factors which are of influence upon natural phenomena, and who is not guided by a theory, with the work of an engineer, who wishes to connect two towns, situated on the opposite sides of a range of mountains, by means of a tunnel, and who breaks down a whole mountain in order to be sure of finding the shortest and easiest route. The scientific worker who uses a theory forms an opinion of the most convenient way of solving his problems. He is to be compared with the engineer who, by means of a preliminary survey, has formed an opinion of the relative positions of the two towns and of the mountain range, and with the aid of his instruments points out the direction on the two sides of the mountain in which his men have to bore the tunnel. If the opinion of the engineer is not quite correct, the tunnels from both sides will not meet precisely in the middle of the mountain, but the axis of the one will be distant some few feet from that of the other. Small corrections have then to be made at the point of intersection by widening the tunnel.

In the same way the theoretical considerations of the scientific worker may lead him a little astray; they must then be corrected slightly and improved. But from his quantitative measurements he finds the natural law which connects together the two factors examined. For building up his theory, the investigator has generally to carry out a series of preliminary measurements, in order to obtain a guiding hypothesis.

From what we have said, the extremely high value of theories will be clear. But theories should always be regarded only as a help, as an instrument or tool. Many, who are not acquainted with theoretical work, express the opinion that a theory should have the character of absolute truth and certainty. They say, for instance, that the emission theory of light has no value, because it has been replaced by the vibration theory of light; and when they hear that this theory also has lost its supremacy, being replaced by the electromagnetic theory of light, they shake their heads, and intimate that it would have been better to have saved the time taken up in working out theories, and to have employed it in making experiments. Such persons fail to appreciate the fruitful use that Newton made of the emission theory, and the wonderful development of optics by means of the undulation theory. The opinion of these men is as well founded as that of a workman who rejects the use of tools because these will be replaced sooner or later by more refined machines.

Very often we hear the opinion expressed that a theory has little or no value, because it may be possible to build up another theory on other foundations. This is just as prudent as to recommend the giving up of an instrument in our possession because it perhaps might be possible to construct a better instrument out of another material, before the latter has been made and shown to do more or better work than the old one.

We have used the old method of comparing a theory with a tool or an instrument. One may then ask, "How should an hypothesis be represented according to this analogy?" An hypothesis may be compared with an instrument the name of which is characterized by the suffix "scope," *e.g.* electroscope, a theory being then compared with an instrument the name of which has the suffix "meter," *e.g.* electrometer. An electroscope enables us to decide if a body is charged with free electricity, and nothing more. The hypothesis has only a qualitative meaning, and tells us nothing about the quantity. Now, one may so elaborate an electroscope that an electrometer is obtained, *e.g.* by fixing a scale behind the gold leaves of the electroscope. We have then the electrometer of Hankel or of Exner. These are from an absolute standpoint somewhat imperfect electrometers, but they serve excellently for many purposes. The absolute electrometer and the quadrant-electrometer of Lord Kelvin may also be regarded as highly perfected elaborations of the gold-leaf electroscope.

In the same manner it is generally possible to work up an hypothesis into a theory by the help of quantitative measurement. Using the new theory, we make a large number of measurements in the new field and discover new connections between the various factors, or new rules that lead to new conceptions, *i.e.* to new hypotheses. These, on the other hand, may thereafter be developed into new theories, which are connected with the old theory out of which they have arisen. These new theories may then be regarded as branches of the old one, and by working further in the same direction we elaborate a system of minor theories, which are all connected with the dominant theory. It is this development of theoretical work that characterizes modern exact science. As an example of such a theory which has developed an almost unlimited number of branches in different parts of physical and chemical science, we might cite the theory of the equivalence of different kinds of energy, or, as it is commonly called, the mechanical theory of heat.

The original hypothesis was in this case extremely simple, and may be stated as follows: "it is impossible to obtain perpetual motion, *i.e.* to construct a machine which performs work without consuming energy of any kind."

In Carnot's hands we see the first wonderful development of this hypothesis into a grand theory. This engineer officer had a practical end in view in his researches—he wished to develop a theory of the engines which transform heat into work, and which at that time had an ever-increasing importance. This work was forgotten, however, and we see how Mayer used the same hypothesis as Carnot, and worked out a theory of equivalence of heat and mechanical work. The work of Mayer was despised, and only a few scientists took notice of it. The same was the case with the simultaneous work of a Danish engineer Colding. But after these pioneers came a procession of theoretical philosophers, before all Joule, Helmholtz, Kelvin, Maxwell, and Clausius, who developed the theories of Carnot and Mayer, so that a theoretical system was built up which now dominates the whole of physical and a great part of chemical science. The greatest modern developments in this system are due to van't Hoff and Gibbs as far as chemical phenomena are concerned, and to Bartoli, Boltzmann, Wien, and Planck in the case of optical phenomena.

No other work has so strongly impressed its character upon modern science as this theoretical development. It is therefore wholly incomprehensible how the opinion can be expressed and find admiring auditors, that it would be better not to work theoretically, but only to collect and register experiments. This opinion may be regarded as a repetition of Rousseau's maxim, that the uncivilized condition is the best for humanity.

In ordinary research work the *rôle* played by fancy is often very limited. Nearly all the attention is at first given to the exact determination of the experimental groundwork. When a sufficient number of data has been collected the theoretical development begins. Then the main point is to combine the results of the different experiments, finding out the common and general features, but leaving out those occasional peculiarities occurring in every experiment owing to unavoidable errors. In this work fancy plays the most important *rôle*, in that from an infinite number of possible combinations it must choose the most simple and most probable. Often the experiments are not exact enough, so that the occasional deviations wholly conceal the law that we seek. We may then use two different methods of procedure. The more rational one is to improve the methods of working, and especially to construct very accurate instruments, so that the experimental errors are reduced to a low value. But sometimes we cannot do this, especially in statistical work. We must then increase the number of observations. If we record forty observations, the experimental error of the mean is only half as great as that of the mean of ten observations, and that of a thousand observations is only the tenth part of that of ten. The experimental error of the mean is in general inversely proportional to the square root of the number of observations. As is evident, it is very laborious to reduce the experimental errors in this way, but sometimes it is the only practical method. If by these means we have found a real law, there will be a regularity in the deviations of the observed mean values from those calculated by its help, so that these differences decrease proportionally to the square root of the number of observations.

Some laws of Nature seem to be exactly true. Thus, for instance, experiments have been made to test the validity of Ohm's law within the widest possible limits, by changing the intensity of the current from the lowest value that we are able to measure exactly with the help of galvanometers, to the highest that we can obtain with the help of modern electrical machines. The result has been that no deviation, greater than the experimental errors, could be detected. Other such laws are, for instance: the law of Newton for the attraction between material bodies; the corresponding law of Coulomb for electrical quantities; Snell's law of the refraction of light; and the two laws of thermodynamics.

But there is another class of laws, and these seem to be the more common; they are only valid within certain limits. As an example of such a law, that of Boyle or Mariotte may be considered. This law asserts that, if we enclose a given quantity of gaseous matter, for instance oxygen, in a vessel of variable volume (such as a cylinder in which works a piston), then the pressure of the gas is inversely proportional to the volume of the vessel. This is true for very low pressures, as Rayleigh[2] has demonstrated recently, but at higher pressures noticeable deviations occur. The first researches on Boyle's law did not lead to the observation of these deviations; hence it was thought that the law was exact. Further researches, carried out in the beginning of the nineteenth century by French savants, showed deviations, but these investigators were so convinced of the exactness of Boyle's law, that they overlooked the rather insignificant deviations which they had observed. As we now examine their experimental results, we find that nearly all the observed deviations were in the same direction, namely, that the observed pressure at small volumes was lower than that calculated. The nature of experimental errors is, however, that they fall as much in the one direction as in the other. Errors that fall in one direction only, are called systematic errors, and they are due to a definite cause, working always in the same direction. As now no reason for the occurrence of systematic errors can be detected in the present case, the investigators of the law of Boyle might have drawn the conclusion that it ceases to be exact at higher pressures. More accurate and wide-reaching work, especially that of Regnault, Natterer, and Amagat, showed well-marked regularities for these deviations. In this way scientists were led to seek for a new law, which agreed with Boyle's law at extremely low pressures and on the other hand showed no deviations from the experimental result at high pressures. Such a law was discovered by van de Waals, and it gave a startling insight into the physical nature of the gases.

But still further research made evident that the law of van de Waals does not cover all cases. With increasing experimental work, the general features of the deviations from van de Waals' law will probably soon be understood sufficiently for the formulation of a new law, by means of which we shall gain still more mastery over Nature than we now possess.

In this way theoretical work proceeds step by step, increasing our knowledge of Nature. If we have found a law which like that of Boyle, is only valid within certain limits, it would be very unreasonable to say that it had no merit at all. Many of the wonderful applications of thermodynamics are based upon the assumption of the validity of Boyle's law. We have to seek the probable causes of the insufficiency of the law in question where the deviations are found, and to apply the necessary corrections, which will with time show themselves to be subject to another law, which completes the original law. In this manner scientific progress is made. The theoretics considerations indicate in which direction further work may be done with the greatest effect. Without them the advance would be very slow, or even impossible.

But the value of theoretical work is generally not limited to that particular part of science which is the direct object of the work in question. The example of Ohm's law shows us that the work of Fourier on the conduction of heat could, with little variation, be adapted to the explanation of the conduction of electricity. But the same law is, with suitable adaptation also valid for the diffusion of dissolved matter, *e.g.* salts in solvent, water. Here the dissolved substance flows from places of higher to those of lower concentration, just as heat flows from places of higher to those of lower temperature, or as electricity flows from places of higher to those of lower potential. The corresponding law of diffusion is called Fick's law, after its discoverer, and it has precisely the same form as that of Fourier or of Ohm.

Further, the benefit of theoretical researches is not confined to the science in which they are carried out. Mathematics was already developed to a high degree in the early ages, and astronomy could use the results of that science, and in this way develop very early. The mathematical and astronomical knowledge of Galileo fitted him to lay the foundations of rational mechanics. Physics is a much more wide-reaching science than mechanics; in its own domain it utilizes the results of the work done in the latter science (*e.g.* the theory of potential), and has thus developed after the same plan. Chemistry is a still more extensive science than physics; it has adapted the laws of thermodynamics to explain phenomena in chemistry, and in the last quarter of a century has thereby won for itself a theoretical basis of the most wide-reaching character. In the same way the biological sciences, which have to do with much more complex phenomena than chemistry, have taken over modern theoretical methods from chemistry, and have in recent times made progress of the most promising character.

The solidity of the different branches of natural science is nowhere so evident as in the domain of theoretical research. All these sciences are highly interested in the theoretical work now going on in chemistry, which is not only raising this science to a higher scientific level, but also giving us a well-founded hope of increasing progress in neighbouring sciences, and especially in biology, which has the greatest importance for the development of humanity.

Notes

1 Cf. Ohm, *Die galvanische Kette,* Neudruck. Leipzig and Wien: Toeplitz and Deuticke. 1887.

2 Rayleigh, *Trans. Roy. Soc.,* 1901, **196,** 205; *Proc. Roy. Soc.,* 1905, **74,** 446; *Phil. Trans.,* A., 1905, **204,** 351.

Editorial Headnote

Svante Arrhenius, *Theories of Solution* (New Haven: Yale University Press, 1912), xvii–xx

Emphasises the importance of theoretical demonstration and its status viz-a-viz experimental demonstration in physical chemistry.

21

THEORIES OF SOLUTION

Svante Arrhenius

Source: New Haven: Yale University Press, 1912, xvii–xx

It is an exquisite honor to speak from this platform in this celebrated university where Willard Gibbs enunciated his standard work on "the theory of heterogeneous equilibria." I also feel very much indebted for the invitation to give a series of Silliman lectures, which have been delivered by the most prominent men of the scientific world. I therefore extend to you my warmest thanks for having conferred this rare distinction upon me.

The object of my lectures will be some chapters of modern physical chemistry. This branch of science has been treated in a rather great number of good or even excellent text-books, here as well as in Europe. It is therefore a difficult task to give something new and something which has not already before been worked out in a masterly manner. But it seems to me as imprudent as to carry owls to Athens to give you a course of physical chemistry as it is presented in the text-books. Therefore I have confined myself to some problems which are now under debate and which are still not finished but which promise the greatest interest for further investigations. Also when I refer to older investigations, I try to exhibit such facts as have not in a higher degree attracted the attention of the authors of text-books in this branch, and thereby to give a broader and more complete view of the exceptionally fertile field which we cultivate. I have often found it useful to drag old historical data into the light especially in order to give an idea of the strict harmonical development of this subject, a circumstance which is often forgotten even to such a degree that physical chemistry is generally represented as if it had like Minerva sprung out quite fully developed from Jupiter's head.

What I wish to express to you is therefore something which according to my opinion supplements the excellent text-books which you already have perused. It is of course agreeable to me to lay before you my personal individual ideas. On the other hand it is quite clear that it would be a great mistake if my attempt were understood to be designed to embrace all the branches or even the most important branches of modern physical chemistry. But I feel quite sure that such an intelligent auditory as that to which I have the honor of speaking here will not commit this mistake and I trust the lacunæ of my exposition will be easily filled up by you who have been introduced so thoroughly into the different chapters of physical

DOI: 10.4324/9781003009337-27

chemistry by your excellent scientific leaders and by the unrivalled interest which the enlightened scientific opinion here in America even more than in the old world attaches to this wonderful branch which is called physical chemistry and which is of the greatest use not only for the most important doctrines of the natural sciences and medicine but also for its far-reaching applications to modern industry. There are very few doctrines in exact science where so few lecture experiments are shown as in physical chemistry. This depends upon its theoretical character. The methods of working are taken from the science of physics. There is almost only one chapter, which is so to say specially adapted for lecture experiments in physical chemistry, namely, that regarding catalytic action. For this chapter, a number of demonstrative experiments are worked out by Professor A. A. Noyes and G. V. Sammet in Boston. Another number of good lecture experiments are described in a little book by Professor Emil Baur in Braunschweig. It may suffice to draw attention to these expositions of the relatively small prominence of the experimental part of our science. The great progress in physical chemistry depends upon the quantitative measurements on which the general laws are based. In most cases it has been necessary to collect the evidence from a very vast field and an extraordinarily great number of experiments in order to get a view of the real situation of the problems and of the possibility of solving them. Chemistry works with an enormous number of substances, but cares only for some few of their properties; it is an extensive science. Physics on the other hand works with rather few substances, such as mercury, water, alcohol, glass, air, but analyses the experimental results very thoroughly; it is an intensive science. Physical chemistry is the child of these two sciences; it has inherited the extensive character from chemistry. Upon this depends its all-embracing feature, which has attracted so great admiration. But on the other hand it has its profound quantitative character from the science of physics. From this circumstance the great solidity and strength of our science is derived. The results of these quantitative measurements regarding the properties of a very great number of substances is very difficult to present by lecture experiments; they must be given in diagrams and tables. I feel it necessary to explain from the very beginning why I have preferred to give a series of theoretical lectures rather than of experimental ones; the cause lies in the very character of the modern development of our science. Therefore modern physical chemistry is often called general or theoretical chemistry as in the two excellent German text-books of Ostwald and Nernst.

The theoretical side of physical chemistry is and will probably remain the dominant one; it is by this peculiarity that it has exerted such a great influence upon the neighboring sciences, pure and applied, and on this ground physical chemistry may be regarded as an excellent school of exact reasoning for all students of natural sciences.

Part 5

ASTRONOMY

Editorial Headnote

John F.W. Herschel, *A Treatise on Astronomy* (London: Longman, Rees, Orme, Brown, Green & Longman, 1833), 1–9

On the necessary preparation of the scientific mind.

22

A TREATISE ON ASTRONOMY

John F.W. Herschel

Source: London: Longman, Rees, Orme, Brown, Green & Longman, 1833, 1–9

A TREATISE ON ASTRONOMY

Introduction

(1.) In entering upon any scientific pursuit, one of the student's first endeavours ought to be, to prepare his mind for the reception of truth, by dismissing, or at least loosening his hold on, all such crude and hastily adopted notions respecting the objects and relations he is about to examine as may tend to embarrass or mislead him; and to strengthen himself, by something of an effort and a resolve, for the unprejudiced admission of any conclusion which shall appear to be supported by careful observation and logical argument, even should it prove of a nature adverse to notions he may have previously formed for himself, or taken up, without examination, on the credit of others. Such an effort is, in fact, a commencement of that intellectual discipline which forms one of the most important ends of all science. It is the first movement of approach towards that state of mental purity which alone can fit us for a full and steady perception of moral beauty as well as physical adaptation. It is the "euphrasy and rue" with which we must "purge our sight" before we can receive and contemplate as they are the lineaments of truth and nature.

(2.) There is no science which, more than astronomy, stands in need of such a preparation, or draws more largely on that intellectual liberality which is ready to adopt whatever is demonstrated, or concede whatever is rendered highly probable, however new and uncommon the points of view may be in which objects the most familiar may thereby become placed. Almost all its conclusions stand in open and striking contradiction with those of superficial and vulgar observation, and with what appears to every one, until he has understood and weighed the proofs to the contrary, the most positive evidence of his senses. Thus, the earth on which he stands, and which has served for ages as the unshaken foundation of the firmest structures, either of art or nature, is divested by the astronomer of its attribute of fixity, and conceived by him as turning swiftly on its centre, and at the same time moving onwards through space with great rapidity. The sun and the moon,

DOI: 10.4324/9781003009337-29

which appear to untaught eyes round bodies of no very considerable size, become enlarged in his imagination into vast globes,—the one approaching in magnitude to the earth itself, the other immensely surpassing it. The planets, which appear only as stars somewhat brighter than the rest, are to him spacious, elaborate, and habitable worlds; several of them vastly greater and far more curiously furnished than the earth he inhabits, as there are also others less so; and the stars themselves, properly so called, which to ordinary apprehension present only lucid sparks or brilliant atoms, are to him suns of various and transcendent glory—effulgent centres of life and light to myriads of unseen worlds: so that when, after dilating his thoughts to comprehend the grandeur of those ideas his calculations have called up, and exhausting his imagination and the powers of his language to devise similes and metaphors illustrative of the immensity of the scale on which his universe is constructed, he shrinks back to his native sphere; he finds it, in comparison, a mere point; so lost—even in the minute system to which it belongs—as to be invisible and unsuspected from some of its principal and remoter members.

(3.) There is hardly any thing which sets in a stronger light the inherent power of truth over the mind of man, when opposed by no motives of interest or passion, than the perfect readiness with which all these conclusions are assented to as soon as their evidence is clearly apprehended, and the tenacious hold they acquire over our belief when once admitted. In the conduct, therefore of this volume, we shall take it for granted that our reader is more desirous to learn the system which it is its object to teach as it now stands, than to raise or revive objections against it; and that, in short, he comes to the task with a willing mind; an assumption which will not only save ourselves the trouble of piling argument on argument to convince the sceptical, but will greatly facilitate his actual progress, inasmuch as he will find it at once easier and more satisfactory to pursue from the outset a straight and definite path, than to be constantly stepping aside, involving himself in perplexities and circuits, which, after all, can only terminate in finding himself compelled to adopt our road.

(4.) The method, therefore, we propose to follow is neither strictly the analytic nor the synthetic, but rather such a combination of both, with a leaning to the latter, as may best suit with a *didactic* composition. Our object is not to convince or refute opponents, nor to enquire, under the semblance of an assumed ignorance, for principles of which we are all the time in full possession—but simply to *teach* what we know. The moderate limit of a single volume, and the necessity of being on every point, within that limit, rather diffuse and copious in explanation, as well as the eminently matured and ascertained character of the science itself, render this course both practicable and eligible. Practicable, because there is now no danger of any revolution in astronomy, like those which are daily changing the features of the less advanced sciences, supervening, to destroy all our hypotheses, and throw our statements into confusion. Eligible, because the space to be bestowed, either in combating refuted systems, or in leading the reader forward by slow and measured steps from the known to the unknown, may be more advantageously devoted to such explanatory illustrations as will impress on him a familiar and, as it were, a practical sense of the sequence of phenomena, and the manner in which they are produced. We shall not, then, reject the analytic course where it leads more easily and directly to our objects, or in any way fetter ourselves by a rigid adherence to method. Writing only to be understood, and to communicate as much information in as little space as possible, consistently with its *distinct* and *effectual* communication, we can afford to make no sacrifice to system, to form, or to affectation.

(5.) We shall take for granted, from the outset, the Copernican system of the world; relying on the easy, obvious, and natural explanation it affords of all the phenomena as they come to be described, to impress the student with a sense of its truth, without either the formality of demonstration or the superfluous tedium of eulogy, calling to mind that important remark of Bacon:—"Theoriarum vires, arcta et quasi se mutuo sustinente partium adaptatione, quâ, quasi in orbem cohærent, firmantur;[1]" nor failing, however, to point out to the reader, as occasion offers, the contrast which its superior simplicity offers to the complication of other hypotheses.

(6.) The preliminary knowledge which it is desirable that the student should possess, in order for the more advantageous perusal of the following pages, consists in the familiar practice of decimal and sexagesimal arithmetic; some moderate acquaintance with geometry and trigonometry, both plane and spherical; the elementary principles of mechanics; and enough of optics to understand the construction and use of the telescope, and some other of the simpler instruments. For the acquisition of these we may refer him to those other parts of this Cyclopædia which profess to treat of the several subjects in question. Of course, the more of such knowledge he brings to the perusal, the easier will be his progress, and the more complete the information gained; but we shall endeavour in every case, as far as it can be done without a sacrifice of clearness, and of that useful brevity which consists in the absence of prolixity and episode, to render what we have to say as independent of other books as possible.

(7.) After all, we must distinctly caution such of our readers as may commence and terminate their astronomical studies with the present work (though of such,—at least in the latter predicament,—we trust the number will be few), that its utmost pretension is to place them on the threshold of this particular wing of the temple of Science, or rather on an eminence exterior to it, whence they may obtain something like a general notion of its structure; or, at most, to give those who may wish to enter, a ground-plan of its accesses, and put them in possession of the pass-word. Admission to its sanctuary, and to the privileges and feelings of a votary, is only to be gained by one means,—*a sound and sufficient knowledge of mathematics, the great instrument of all exact enquiry, without which no man can ever make such advances in this or any other of the higher departments of science, as can entitle him to form an independent opinion on any subject of discussion within their range*. It is not without an effort that those who possess this knowledge can communicate on such subjects with those who do not, and adapt their language and their illustrations to the necessities of such an intercourse. Propositions which to the one are almost identical, are theorems of import and difficulty to the other; nor is their evidence presented in the same way to the mind of each. In teaching such propositions, under such circumstances, the appeal has to be made, not to the pure and abstract reason, but to the sense of analogy,—to practice and experience: principles and modes of action have to be established, not by direct argument from acknowledged axioms, but by bringing forward and dwelling on simple and familiar instances in which the same principles and the same or similar modes of action take place; thus erecting, as it were, in each particular case, a separate induction, and constructing at each step a little body of science to meet its exigencies. The difference is that of pioneering a road through an untraversed country and advancing at ease along a broad and beaten highway; that is to say, if we are determined to make ourselves distinctly understood, and will appeal to reason at all. As for the method of *assertion*, or a direct demand on the *faith* of the student (though in some complex cases indispensable, where illustrative explanation would defeat its own end by becoming tedious and burdensome to both parties), it is one which we shall neither adopt ourselves nor would recommend to others.

(8.) On the other hand, although it is something new to abandon the road of mathematical demonstration in the treatment of subjects susceptible of it, and teach any considerable branch of science entirely or chiefly by the way of illustration and familiar parallels, it is yet not impossible that those who are already well acquainted with our subject, and whose knowledge has been acquired by that confessedly higher and better practice which is incompatible with the avowed objects of the present work, may yet find their account in its perusal,—for this reason, that it is always of advantage to present any given body of knowledge to the mind in as great a variety of different lights as possible. It is a property of illustrations of this kind to strike no two minds in the same manner, or with the same force; because no two minds are stored with the same images, or have acquired their notions of them by similar habits. Accordingly, it may very well happen, that a proposition, even to one best acquainted with it, may be placed not merely in a new and uncommon, but in a more impressive and satisfactory light by such a course—some obscurity may be dissipated, some inward misgiving cleared up, or even some link supplied which may lead to the perception of connections and deductions altogether unknown before. And the probability of this is increased when, as in the present instance, the illustrations chosen have not been studiously selected from books, but are such as have presented themselves freely to the author's mind as being most in harmony with his own views; by which, of course, he means to lay no claim to originality in all or any of them beyond what they may really possess.

(9.) Besides, there are cases in the application of mechanical principles with which the mathematical student is but too familiar, where, when the data are before him, and the numerical and geometrical relations of his problems all clear to his conception,—when his forces are estimated and his lines measured,—nay, when even he has followed up the application of his technical processes, and fairly arrived at his conclusion,—there is still something wanting in his mind—not in the evidence, for he has examined each link, and finds the chain complete—not in the principles, for those he well knows are too firmly established to be shaken—but precisely in the *mode of action*. He has followed out a train of reasoning by logical and technical rules, but the signs he has employed are not pictures of nature, or have lost their original meaning as such to his mind: he has not seen, as it were, the process of nature passing under his eye in an instant of time, and presented as a whole to his imagination. A familiar parallel, or an illustration drawn from some artificial or natural process, of which he has that direct and individual impression which gives it a reality and associates it with a name, will, in almost every such case, supply in a moment this deficient feature, will convert all his symbols into real pictures, and infuse an animated meaning into what was before a lifeless succession of words and signs. We cannot, indeed, always promise ourselves to attain this degree of vividness in our illustrations, nor are the points to be elucidated themselves always capable of being so *paraphrased* (if we may use the expression) by any single instance adducible in the ordinary course of experience; but the object will at least be kept in view; and, as we are very conscious of having, in making such attempts gained for ourselves much clearer views of several of the more concealed effects of planetary perturbation than we had acquired by their mathematical investigation in detail, we may reasonably hope that the endeavour will not always be unattended with a similar success in others.

(10.) From what has been said, it will be evident that our aim is not to offer to the public a technical treatise, in which the student of practical or theoretical astronomy shall find consigned the minute description of methods of observation, or the formulæ he requires prepared to his hand, or their demonstrations drawn out in detail. In all these the present work will be found meagre, and quite inadequate to his wants. Its aim is entirely different; being to present in each case the mere ultimate *rationale* of facts, arguments, and processes; and, in all cases of mathematical application, avoiding whatever would tend to encumber its pages with algebraic or geometrical symbols, to place under his inspection that central thread of common sense on which the pearls of analytical research are invariably strung; but which, by the attention the latter claim for themselves, is often concealed from the eye of the gazer, and not always disposed in the straightest and most convenient form to follow by those who string them. This is no fault of those who have conducted the enquiries to which we allude. The contention of mind for which they call is enormous; and it may, perhaps, be owing to their experience of *how little* can be accomplished in carrying such processes on to their conclusion, by mere ordinary *clearness of head;* and how necessary it often is to pay more attention to the purely mathematical conditions which ensure success,—the hooks-and-eyes of their equations and series,—than to those which enchain causes with their effects, and both with the human reason,—that we must attribute something of that indistinctness of view which is often complained of as a grievance by the earnest student, and still more commonly ascribed ironically to the native cloudiness of an atmosphere too sublime for vulgar comprehension. We think we shall render good service to both classes of readers, by dissipating, so far as our power lies, that accidental obscurity, and by showing ordinary untutored comprehension clearly what it *can*, and what it *cannot*, hope to attain.

Note

1 The confirmation of theories relies on the compact adaptation of their parts, by which, like those of an arch or dome, they mutually sustain each other, and form a coherent whole.

Editorial Headnote

John F.W. Herschel, *Results of Astronomical Observations Made During the Years 1834, 5, 6, 7, 8, at the Cape of Good Hope* (London: Smith, Elder and Co., 1847), v–xx

On the laborious methods of preparing necessary scientific equipment. Above all, the following account lays bare the amount of practical preparation required prior to being able to even look, let alone see.

23

RESULTS OF ASTRONOMICAL OBSERVATIONS MADE DURING THE YEARS 1834, 5, 6, 7, 8, AT THE CAPE OF GOOD HOPE

John F.W. Herschel

Source: London: Smith, Elder and Co., 1847, v–xx

INTRODUCTION

(i.) THE work which is now presented to the astronomical public, completes a review of the sidereal heavens, which I commenced about the year 1825, proposing to myself, at that time, no further object than a re-examination of the nebulæ and clusters of stars discovered by my Father in his "Sweeps of the Heavens," and described by him in three catalogues presented to the Royal Society, and published by that illustrious body in their Transactions for the years 1786, 1789, and 1802. This re-examination occupied about eight years, and its results were presented to the Royal Society in the year 1833, in the form of a catalogue, arranged in order of Right Ascension, and published in their Transactions for the same year. In this work are recorded observations of 2306 nebulæ and clusters, of which 1781 are identical with objects occurring, either in my Father's catalogues, in the small but interesting collection published by Messier in the *Mémoires de l'Académie des Sciences* for 1771, and the *Connaissance des Tems* for 1783, 1784; and in M. Struve's catalogue of double stars: the remaining 525 are new. Besides these more especial objects of pursuit, however, a great number of double stars, of all classes and orders, were noticed and described, and their places taken, to the amount, altogether, of between three and four thousand, the observations of which, being reduced and arranged in order of Right Ascension, have been, from time to time, published by the Royal Astronomical Society of London, in six catalogues, which will be found in the 2nd, 3rd, 4th, 6th, and 9th volumes of their Transactions.

(ii.) Having so far succeeded to my wish (the places of the objects thus determined proving, on the whole, satisfactory), and having by this practice acquired sufficient mastery of the instrument employed (a reflecting telescope of 18¼ inches clear aperture, and 20-feet focus, on my Father's construction), and of the

DOI: 10.4324/9781003009337-30

delicate process of polishing the specula; being, moreover, strongly invited by the peculiar interest of the subject, and the wonderful nature of the objects which presented themselves in the course of its prosecution, I resolved to attempt the completion of a survey of the whole surface of the heavens, and for this purpose to transport into the other hemisphere the same instrument which had been employed in this, so as to give a unity to the results of both portions of the survey, and to render them comparable with each other.

(iii.) Accordingly, having placed the instrument in question, as well as an equatorially mounted achromatic telescope of five inches aperture, and seven feet focal length, by Tulley, which had served me for the measurement of double stars in England, together with such other astronomical apparatus as I possessed, in a fitting condition for the work, and taken every precaution, by secure packing, to insure their safe arrival in an effective state, at their destination, they were conveyed (principally by water carriage) to London, and there shipped on board the *Mount Stewart Elphinstone*, an East India Company's ship, ——Richardson, Esq. Commander, in which, having taken passage for myself and family for the Cape of Good Hope, we joined company at Portsmouth, and sailing thence on the 13th November, 1833, arrived, by the blessing of Providence, safely in Table Bay, on the 15th January, 1834, and landed the next morning, after a pleasant voyage, diversified by few nautical incidents, and without seeing land in the interim. It was most fortunate that, availing himself of a very brief opportunity afforded by a favourable change of wind, our captain put to sea when he did, as we subsequently heard that, immediately after our leaving Portsmouth, and getting out to sea, an awful hurricane had occurred from the S. W. (of which we experienced nothing), followed by a series of south-west gales, which prevented any vessel sailing for six weeks. In effect, the first arrival from England, after our own, was that of the *Claudine*, on the 4th of April, with letters dated January 1st.

(iv.) Having disembarked the instruments without accident, and placed them, temporarily, in one of the Government storehouses (being permitted to do so through the obliging kindness of W. Petrie, Esq. Deputy Commissary General of the Colony), my next care was to look out for a comfortable residence in a locality suitable for their erection. This I was fortunate enough speedily to find at the mansion of a Dutch proprietor, W. A. Schonnberg, Esq., bearing the name of Feldhuysen, or Feldhausen, about six miles from Cape Town, in the direction of Wynberg, a spot charmingly situated on the last gentle slope at the base of the Table Mountain, on its eastern side, well sheltered from dust, and, as far as possible from wind, by an exuberant growth of oak and fir timber; far enough removed from the mountain to be, for the most part, out of the reach of annoyance from the clouds which form so copiously over and around its summit, yet not so far as to lose the advantage of the reaction of its mural precipices against the south-east winds which prevail with great violence during the finer and clearer months, but which seldom *blow home* to the rock on this side, being, as it were, gradually heaved up by a mass of comparatively quiescent air imprisoned at the root of the precipice, and so gliding up an inclined plane to the summit on the windward side, while they rush perpendicularly down on the leeward with tremendous violence like a cataract, sweeping the face of the cliffs towards Cape Town, which they fill (as well as the valley in which it stands) with dust and uproar, chiefly during the night. This residence needing some repairs, and being, in fact, not then actually vacant, a temporary residence was secured, at a convenient distance, in a tenement belonging to ——Borcherds, Esq., of Cape Town, called Welterfrieden, whence I could superintend the erection of the instruments, which was forthwith commenced, and pushed forward with such effect, that on the 22nd of February I was enabled to gratify my curiosity by a view of α Crucis, the nebula about η Argûs, and some other remarkable objects, in the 20-feet reflector; and, on the night of the 5th of March, to commence a regular course of sweeping.

(v.) Shortly after, the erection of a building was commenced to receive the equatorial instrument, which, so soon as the walls were settled, and the pier, which was of brick, laid in Roman cement, consolidated, was placed on its supports, and being surmounted by a revolving roof of a peculiar construction (See Art. 158), contrived by myself, and constructed in England for the purpose, was brought into approximate adjustment; and on the 2nd of May, a series of micrometrical measures of southern double stars was commenced by the measurement of α Centauri, the chief among them. I ought to observe that, on unpacking this and the other instruments and apparatus, not a single article was found to have sustained injury.

(vi.) The Frontispiece to this volume exhibits the Reflector, and the building containing the Equatorial, in their enclosure, a kind of orchard, surrounded on all sides by trees, but commanding a tolerably near approach to both the southern, eastern, and northern horizon. To the west, the nearer vicinity of the trees, though cutting off much of the sky, yet afforded a valuable protection from the fury of the north-west gales which prevail in the winter months. Nor was it without some regret on this account, that the removal of a portion of this friendly shelter was resolved on to afford a view of the comets of Encke and Halley, when so situated as to require it. The exact geographical site of Feldhausen was found by trigonometrical measurement (See Appendix E), to be $2' 53''.55$ south, and $0^{m} 4^{s}.11$ west of that of the Royal Observatory of the Cape, that is to say, in Lat. $33° 58' 56''.55$, Long. $22^{h} 46^{m} 9^{s}.11$ from Greenwich. Its altitude above the site of that building was ascertained by barometrical observation to be 112.23 feet, or about 142 feet above the mean sea-level of Table Bay (See Appendix B).

(vii.) As I have mentioned the Royal Observatory of the Cape, I must take this distinct and early opportunity to acknowledge my many and great obligations to my excellent friend, Thomas Maclear, Esq., its director, whose ardent love for the science to which he has devoted himself, seconded by talent of no common order, by resource which no difficulties exhaust, and by activity which no exertion wearies, has secured, and continues to secure, for the noble establishment under his direction, a reputation of the highest rank, and for himself, the best titles to the gratitude of his country, and the approbation of the government he serves. To his kindness I am indebted for much and most valuable aid in fixing, with all the precision which the fine instruments under his command could afford, the Right

Ascensions of a great many stars of which a knowledge was indispensable in the reduction of my sweeps, and which I neither could obtain with sufficient exactness from the existing catalogues, nor had any means of ascertaining for myself, unless by entering upon an express course of observation for that very purpose, such as formed no part of my original design, and for which a much superior transit to that I had brought out with me would have been requisite. Previous to leaving England, I had procured a MS. list of the stars of the Brisbane Catalogue, then preparing for publication, containing the places of between seven and eight thousand southern stars, and on this Catalogue, when it should appear, I had confidently reckoned for as many well determined zero stars as I could possibly need, the system of observing I proposed to follow being entirely differential. With the Polar distances of that catalogue, I seldom had much occasion to find fault; but as respects the Right Ascensions, I experienced less satisfaction in its use. On mentioning this circumstance to Mr. Maclear, he, with ready alacrity, at once offered to determine this important element, by direct observation, for as many zero stars as I might require, thus rescuing me from a very serious difficulty. I need hardly add that I gladly availed myself of, and drew largely on his kindness; and the results of his observations in the cases of about 670 stars, fixed by two or three observations of each, have proved a most material assistance to me in the reduction of my sweeps. Nor was this the only instance of useful and effective aid rendered me from the same quarter.

(viii.) Previous to quitting England, I had engaged the services of an attendant for the purpose of working the sweeping, and other mechanical movements, of the Reflector during the observations, and executing any necessary repairs. John Stone, the person so engaged, to the useful, and, indeed, indispensable qualifications of a ready mechanic, whether in wood or iron work, joined that of experience in this particular employment, having performed that office for me during a considerable portion of my review of the northern heavens, with undeviating steadiness and regularity, as he continued to do during the whole of that of the southern, without once absenting himself from his duty. This indispensable manual aid excepted, it is right to mention that the whole of the observations, as well as the entire work of reducing, arranging, and preparing them for the press, has been executed by myself. The repolishing of the reflectors, it will of course be understood, could be delegated to no other person.

(ix.) Of these I was provided with three, viz., one made by my Father, and used by him in his 20-feet sweeps, and other observations; one made by myself, under his inspection and instructions; and one which I ground and figured subsequently, but which was cast at the same time, and from the same metal as that last mentioned. They are each 18¼ inches of clear diameter of polished surface, and all, so far as I am able to judge, equally reflective when freshly polished, and in every respect similar in their performance. One of them (the first completed, of the two newer ones above mentioned) is about two inches longer in focus than the others, a difference which, the figure being otherwise good, it has not been thought worth while to correct.

(x.) The operation of repolishing was performed whenever needed, the whole of the requisite apparatus being brought for the purpose. It was very much more frequently required than in England; and it may be regarded as fortunate that I did not, as at first proposed (relying on the possession of three perfect metals), leave the apparatus in question behind. Being apprehensive that in a climate so much warmer, difficulties would arise in hitting the proper temper of the polishing material, slight imperfections of surface, induced by exposure, were for a while tolerated; but, confidence in this respect once restored, and practice continually improving, I soon became fastidious, and on detection of the slightest dimness on any part of the surface, the metal was at once remanded to the polisher.

(xi.) And here, perhaps, I may be allowed a digression on a point of the utmost importance in the use of reflecting specula, viz., the mode of supporting the metal in its case. This, in my own practice, is provided for as follows:—between the back of the case and the mirror are interposed six or eight thicknesses of coarse woollen baize, or blanketing, of even texture, and quite free from knots, stitched together at the edges to prevent any hard substance from getting between them. On this bed the metal is laid flat, and being shaken into a concentric situation, as respects the rim of the case, two supports formed of strips of similar woollen stuff, many times doubled, occupying about 30° each of the circumference of the case, are introduced, so as to leave an arc of about 40° unoccupied, opposite the point which is intended to be placed lowermost in the tube. The case being then raised into an inclined position by the other handle, and slightly shaken, the mirror takes its own free bearing on these supports, which by their elasticity obviate the possibility of any lateral compression which might go to the extent of seriously disfiguring the metallic surface, were the whole vertical pressure of the mirror confined to a hard point near the bottom, or even distributed over two or three metallic bearings in the circumference.

(xii.) Simple, and indeed homely as this mode of bedding the speculum may appear, it is, I am satisfied, as effectual (it certainly is quite as little costly and cumbersome) as any which can be contrived. The uniform support of a reflector over its whole extent, is a point of the last importance to its optical performance. *A distortion of figure by flexure, which in the object-glass of a refracting telescope would produce no appreciably injurious effect, would be utterly fatal to distinct vision in a reflecting one.* This will be made apparent if we consider the different modes in which the convergence of the rays is effected by reflexion at a curved metallic surface, and by refraction through one or more glass lenses. In the former case, the whole deviation of the ray from its original direction is effected at one surface, and, in amount, is equal to twice the inclination of that surface to a perpendicular on the original direction. Any alteration, therefore, which flexure may produce on the inclination of the surface (supposed originally parabolic) of a speculum to its axis, at a given point, will produce an angular deviation of the reflected ray from its proper course to a double amount. And the lateral aberration produced by such deviation will be expressed by that double amount of angular flexure, multiplied by the focal distance, and is, therefore, the more injurious the greater is the focal length. On the other hand, in the case of a refracting telescope, the deviation of a ray is effected by the prismaticity of the medium, or media, of which the object-glass consists, at the points where the ray penetrates them, and is independent of the absolute inclination of either separate surface to the axis. In consequence, however, this inclination may change by flexure in any one of the lenses of which the object-glass consists, yet, so long as both surfaces of the lens bend alike, which they cannot but do in any conceivable case of flexure short of fracture, the effective prismatic angle at that point, and consequently, also, the inclination of the transmitted ray to the axis remains absolutely unaltered, and the whole amount of lateral aberration produced is that which arises from the bodily displacement of the transmitting portion of the lens in a direction to or from the focus by the effect of the flexure, a displacement which, in itself utterly inappreciable, is still farther reduced in its effect to produce *lateral* aberration, that it comes to be multiplied by a fraction whose numerator at the maximum is the semi-aperture of the object-glass, and whose denominator is the focal length of it.

(xiii.) This inequality of proportion between the two cases is still farther augmented by the lightness of glass compared with metal, which (surface for surface, and rigidity for rigidity) makes the absolute flexure of a given thickness of glass much less than of an equal thickness of metal. And it is very fortunate that all these conditions hold good; for since there is no possibility of supporting a transparent substance against flexure, without stopping the light, this cause alone (had it acted to the extent it does in reflectors) would long since have banished large object-glasses from practical use. A speculum (I speak from experience) of the dimensions and thickness used in my sweeps, is *totally spoiled* by supporting it on three metallic points at the circumference, when directed to the zenith. The image of every considerable star becomes triangular, throwing out long flaring caustics at the angles. On one occasion, I supported a mirror simply against a flat board, at about 45° elevation from the horizon. In this state its performance was tolerably good; but on stretching a thin packthread vertically down the middle of the board, so as to bring the weight to rest on this as on one axis, the images of stars were elongated, in a horizontal direction, to a preposterous extent, and all distinct vision utterly destroyed by the division of the mirror into two lobes, each retaining something of its parabolic figure, separated by a vertical band, in a state of distortion, and of no figure at all.

(xiv.) Springs are sometimes used by opticians at the back of a small mirror, for the purpose of keeping it home to the rim of its case. They are usually three in number, and so far as mere support of the mirror is concerned, they act as three non-elastic supports would do, *i. e.* disadvantageously. More numerous and weaker springs distributed over the whole of the back of the case would seem to be preferable, but they ought to be so weak that *all* should be *much* compressed and brought into full action by the pressure. And even then, if the metal be confined in its case by a rim in front, a spring-pressure upward, which in a horizontal position of the surface would barely sustain it in contact with the rim, will urge it, when placed vertically against the rim, with a force equal to the whole weight of the metal applied over its surface in a direction perpendicular thereto, and will tend therefore to distort it in that position, to the very same amount, but in an opposite direction, that it would have been distorted by gravity in the other position, had the springs been absent, and the support been given by a metallic ring at the circumference of the back.

(xv.) Nothing, therefore, is gained, so far as obviating flexure is concerned, by a distribution of elastic support, *so long as the mirror is pressed against a rim in front*. To make such a mode of support effective, the metal must be free to rise and fall, and the friction on the lower circumference must be counteracted, either by resting it in a circular iron chair, provided with exterior rollers, or otherwise equably supporting the lower portion of its edge, so as to allow of its shifting forwards or backwards without sensible friction; as, for instance, by suspending the metal from the upper side of its case, in a jointed frame, composed of two semicircles of steel loosely riveted at the extremities of their common, horizontal diameter, to admit of free play at the joint, the lower semicircle being thin, and lined with velvet or flannel to ensure equal pressure against the lower edge of the metal; the upper stronger, and increasing gradually in stiffness towards the top, by which the whole should be suspended on a hinge-pin.[1]

(xvi.) As an experiment, I constructed a case, having the back of stout wood set over with ninety steel springs, of such strength as to be all of them pressed nearly into contact with the wood by the weight of the mirror laid horizontally on them, and distributed with as much uniformity as possible in their points of bearing over the whole surface of the back, the face being left entirely unconfined. So supported, the performance of the mirror was good at all altitudes, but by no means better than when bedded on woollen cloths, as above described, nor, indeed, equal to it; and, in fact, when we consider that each fibre of wool is a delicate coiled spring of almost perfect elasticity, it is clear that no artificial arrangement of metallic springs we can make, can attain the perfection of such a natural one (if we may so apply the term), either in uniformity of distribution, or in delicacy of application. This consideration led me to abandon the use of a spring case, after making a very few sweeps with it, and re-adopt, in its stead, the woollen bedding which had been laid aside to make way for what proved to be, in effect, no substantial improvement on it. It is essential, however, that a great many thicknesses of the baize or blanket employed, should be used, *by which only the effect of flexure in the wooden back itself of the case can be eliminated*. And to keep up the elasticity of the fibre, it should be occasionally taken out and beaten.[2]

(xvii.) A serious, indeed a fatal objection, would appear to lie against the use of flexible, elastic, or any kind of moveable support for the mirror; in the fluctuations liable to be caused by it in the line of collimation when the inclination of the mirror to the horizon is varied, and which must inevitably take place if there be either the least friction of the lower edge on its support, or anything short of mathematical exactness in the distribution of the pressure. But this objection is completely obviated in my mode of observation, by the use of an interior collimating telescope, as described in Phil. Trans. 1833, p. 488. The collimator used in the greater part of these observations differed only from that there described, 1st, in being a more powerful telescope (viz., an achromatic of four feet focal length, with a good object-glass, 2¾ in. aperture), and, 2nd, in. having the plane speculum placed, not as there described, between the object-glass and its focus, but beyond the focus (in which the collimating cross is fixed), so as to illuminate that cross by the light of a lamp external to the tube of the large reflector, collected by a lens upon the cross, both great improvements in practice.

(xviii.) Sweeping (for so I shall continue to denominate the system of observation in zones of 3° breadth in Polar distance in search of new objects), was prosecuted, in the absence of the moon, on all occasions when weather permitted, and the definition of the stars was such as to render it worth while to do so. And this leads me to speak of the climate of the Cape as regards its favourableness or unfavourableness for astronomical observation. In the hot season (from October to March), and especially during the hotter months of that season, the nights are for the most part, superb, at least in all the flat region at a few miles distance from the mountains; but in their immediate vicinity, the south-east wind (then generally prevalent) frequently brings with it a belt of cloud extending many miles from the hills, and cutting off the view of the sky. This (which is sometimes called a black south-easter) comes on often at night, and lasts several nights in succession. Very often, too, when no such impediment exists, the excessive heat and dryness of the sandy plains gives rise to a disturbance of the optical tranquillity of the air, so as to destroy, or much impair, distinct vision, and that frequently in a very singular manner. In some cases, the images of the stars are violently dilated, and converted into ill-defined nebulous balls, or puffs, of 10 or 15″ or more in diameter.[3] In others, they form soft, quiet, round pellets of 3 or 4″ diameter, very unlike the "spurious discs" which they present when best defined, and rather resembling planetary nebulæ. In other cases, again, the structure, as it were, of these pellets is disclosed, and they are seen to arise from an infinitely rapid vibratory movement of the central point, in all possible directions, while, on a few occasions, the appearances have been exceedingly perplexing and singular, and such as there is difficulty in accounting for on any optical principle whatever.[4]

(xix.) Even in the hottest season, however, nights of admirable definition occur, especially looking southwards. But, what is not a little remarkable, in the very hottest *days*, looking *northwards* over the burning tract intervening between Feldhausen and Table, or Saldanha Bay, the most admirable and tranquil definition of the solar spots, and other phenomena of the sun's disc, is by no means unfrequent. In such cases, I presume the strongly heated stratum of air incumbent on the surface of the soil, is swept off by the south-east wind blowing from False to Table Bay, before it ascends high enough to interfere with the visual ray. To how high a temperature the soil is occasionally heated, will be seen in Appendix (C).

(xx.) During the autumnal months of May, June, and July, when the weather is generally cool, and the nights cold, hot winds frequently set in from the north, usually in the night; at first moderate, and alternating with cold gusts from the opposite quarter, but rapidly gaining the prevalence, till, in a few hours, they attain the violence of a hard gale and glowing oven-like heat.[5] Thus they continue to blow, frequently for thirty-six or forty-eight hours, becoming at length cool, and finally settling into rain. Now it is not a little remarkable, that however unfavourable to vision this conflict of hot and cold currents is at the coming on of such a gale, no sooner does it become established than the images become concentrated, and settle down not unfrequently into excellent definition.

(xxi.) It is, however, in the cooler months, from May to October inclusive, and more especially in June and July, that the finest opportunities occur. The state of the air in these months, as regards definition, is habitually good, and imperfect vision is rather the exception than the rule. The best nights occur after the heavy rains which fall at this season have ceased for a day or two; and on these occasions, the tranquillity of the images, and sharpness of vision is such, that hardly any limit is set to magnifying power but what the aberrations of the specula necessitate.[6]

(xxii.) Among the irregular and accidental optical effects of peculiar atmospheric conditions incident to the climate, there are one or two which seem deserving of especial notice. The first is that phenomenon which, when it occurs, I have designated by the epithet, the "nebulous haze." Its effect is to convert every star of the 9th magnitude and upwards, into a "nebulous star," meaning thereby a well-defined star, with a faint, nebulous photosphere of greater or less extent, according to the brightness of the star, surrounding it. This phenomenon occurs in a perfectly clear sky, free from the slightest suspicion of cloud. It comes on very suddenly and unexpectedly, and goes off as suddenly, lasting sometimes only a few minutes; at others, longer. Thus, in sweep 500, Oct. 5, 1834, it commenced at $22^h\ 4^m$ ST, when a star 7 m was observed to be surrounded with it, having come on quite suddenly, and continued to affect all the brighter stars until $22^h\ 54^m$, when it was quite gone, being described as extraordinary in intensity, and very troublesome during its continuance. From this time till $0^h\ 27^m$, all was clear, when it suddenly came on again, "in an instant. A star 7 m was quite free, but on drawing it back" (after it had left the field for re-examination), "it was found to be completely involved," the sky continuing all the while pure, so far as the naked eye could discern. Again, in sweep 598, June 18, 1835, we have "$15^h\ 37^m$ ST. A nebulous haze came on in an instant, extending to stars 9 m; yet the sky is as clear as ever, and the calm unbroken.—"$15^h\ 44^m$—(a star 6 m in the field). The nebulous haze is gone; it did not last two minutes."—"$16^h\ 23^m$. The nebulous haze came on again in a moment." Such remarks might lead to a suspicion of dew upon the eye-piece, or the breath of the observer settling on the glass; but repeated examination (the phenomenon being very common) has satisfied me that such is not the cause, but that it is really of atmospheric origin. Similar nebulous affections occur in our English climate; but it is their much greater frequency, and the suddenness of their appearance and disappearance, which forms so remarkable a feature at the Cape.

(xxiii.) Another peculiarity which has frequently given rise to remark, is the opacity of cloud, as compared with what prevails in England. My Sweeping Registers, at Slough, are full of instances of double stars with small companions, clusters, &c. seen through cloud of considerable apparent density to the unassisted eye. Of this I have elsewhere stated remarkable instances. It was my usual practice there to continue sweeping during moderately-clouded intervals, for the sake of securing at least such new double stars as might occur, or obtaining micrometric measures of known ones placed on the working list. Such advantage could seldom be taken of clouded intervals at Feldhausen, and from the whole tenor of my experience of that locality, the impression remains of a very decided difference, in this respect, between the two stations.

(xxiv.) Meteorological observations of the barometer, thermometers, wet and dry, the actinometer, &c., with the usual record of weather, were made on the appointed *term-days*, or days of equinox and solstice; at first during thirty-six hours, subsequently, during twenty-four only, from hour to hour. These, however, are only valuable when taken in conjunction with others of the same kind; and having already been published in the pages of the *Athenæum*, and elsewhere, and received as much discussion as they have appeared to need, in a report which I had the honour to communicate to the British Association, for the advancement of science, and which has been published by that body, it has been thought unnecessary to reprint them.

(xxv.) It was, however, my intention to have placed on record, in this volume, the results of a great mass of Actinometric observation made since the year 1824 (at which epoch my attention was first directed to that method of ascertaining the intensity of solar radiation, which may not inaptly be termed dynamical, in contradistinction to the usual statical method by the observation of blackened thermometers, photometers, &c.) In addition to these, between five and six hundred sets of such observations obtained at the Cape, would, I supposed, have afforded valuable climatological data, and led to conclusions otherwise important. Unfortunately, the necessity of a correction by a variable factor depending on the temperature attained by the liquid enclosed in the cylinder of the instrument, had not been foreseen and provided for in any of these observations, and only became apparent when the omission was beyond remedy, viz., when the results were drawn out, and ready for press. They are, therefore, suppressed for the present, leaving it to future examination to determine whether among them there may not be a portion, made under such circumstances, or registered in such a manner as to admit of the temperature and, therefore, the correction in question being ascertained and applied, and providing against a similar evil in future by a simple and easily applied addition to the instrument itself.

(xxvi.) The principles upon which the reduction of the observations made with the reflector is executed, are explained at large in the Appendix to my Catalogue of northern nebulæ, Phil. Trans. 1833, p. 482, et seq., and need not here be repeated. Only as respects the precession in Right Ascension, a slight change is introduced. That correction depending both on the Right Ascension and Polar distance of the object, would introduce, if attempted to be included in the general system of interpolation there adopted, a term of double entry which, however, since the zone swept is of small breadth, not exceeding 3°, may be resolved into two; the one a term of single entry, depending only on the time and expressing the precession for a constant polar distance, that of the middle of the zone; the other a correction of this for objects occurring out of the middle of the zone. This latter portion is a term of double entry; but it is necessarily of very small magnitude, seldom exceeding a few tenths of a second of time. In the system of reduction adopted in my former catalogue, this term is simply neglected; or the precession is regarded as the same throughout the breadth of the zone. Thus it becomes a mere function of the time of observation, and merges in the general correction for the time, to be interpolated as there explained. In the reduction of the present series of observations, this did not appear admissible, and it became necessary, therefore, to proceed somewhat differently, making a special exception of the precession in Right Ascension. This was done as follows: all the Right Ascensions of the zero stars were brought up to the beginning of the year nearest to the date of the observation, and the process, in its original form, followed out, as if to reduce the observations of the sweep to that epoch. The reduction (R) in this case, consists of two terms (A) and (B), of single entry, the former a function of the time, the latter of the polar distance. To bring these to the common epoch (1830) adopted throughout, the precession is thrown into the form P + p, where P is the precession for the middle of the zone, and p is the small correction to be applied to P to obtain the precession for any other point out of the middle. Of these, P, being a function of the time, merges in A, and is added to, and so included in it; and p is calculated by a small table, of double entry, and applied to each star according to its proper amount and sign.

(xxvii.) Great and unavoidable delay having occurred in the publication of these pages, an opportunity has been afforded for a careful revision of the Catalogues subsequent to their printing off. In the course of this a considerable list of *corrigenda* has been accumulated, partly arising from error of the press, partly of copying, and partly of calculation. A few nebulæ and double stars also, which are not included in the Catalogues, having escaped the process of arrangement by which they were constructed, have been discovered in the sweeping journals, as well as some additional observations of others which are so included. These are appended accordingly to the list of corrigenda. Some errors which vitiate the letter-press are also noticed. They are few, and for the most part trivial, with the exception of two very obnoxious ones which the reader is requested to bear in mind, or correct for himself. In page 299, the major semiaxis of the orbit of γ Virginis is stated at $9''.69$, whereas it ought to be $3''.58$; and in page 368, equation (A), the numerical coefficient in the value of μ, instead of 3.1514, should be 2744.7.

(xxviii.) It remains to say something as to the mode of introducing this work into the world. To the munificent destination of his Grace the late Duke of Northumberland, of a large sum, in aid of its publication, it owes its appearance as a single and separate work, instead of a series of unconnected memoirs, scattered over the volumes of academical bodies. The lamented decease of that illustrious nobleman prevented his witnessing its final completion. His liberal intentions, however, have been fully carried out by the worthy successor to his titles and his spirit; whose kind and gracious interest in it, I should be wanting in all proper feeling, were I to omit this opportunity of acknowledging.

Notes

1 This was written while ignorant of the very ingenious contrivance by which Lord Rosse affords an equable, or nearly equable support to his large reflector. Comparative trial of the several methods can alone decide, in any proposed case, which is the most effectual.

2 The following memorandum occurs in sweep 687:—

"The mirror, though a capital one, has, for the last three nights, given distorted images with its full aperture, which appear horned and tailed. To-night, before the sweep, I took it out, and carefully smoothed, shook up, and remade its bed. *Now,* nothing can be finer. It lies at its ease, and equally supported all over, and all the appendages are clean gone."

3 During the conjunction of Saturn and γ Virginis, at the end of March, 1834, both were seen in the same field of the 20-feet reflector, in such a state of indefinition, that but for the greater quantity and different colour of the light of Saturn, it could not have been told which was the planet, and which the star. Yet, to all appearance, the night would have been judged a fine one.

4 The discs ill defined and agitated points, surrounded with distant halos. Where thrown out of focus *inwards,* the images, instead of circular discs, would present large central *vacuities,* absolutely void of light, round which a turban-fashioned phenomenon, in a constant state of vorticose motion, was seen. Occasionally, the image so thrown out of focus takes the form of a narrow circular arc, with a terminal concentration; when thrown out of focus *outwards,* a vivid central point, surrounded with an extensive circular area of light. The phenomena have manifestly a reference to the state of the air in the tube of the telescope, and at its aperture, and would seem to indicate the existence of a cone of heated air projecting beyond the aperture of the tube. The tube of a reflector being necessarily open at the mouth, ascending and descending currents of hot and cold air (usually rotating spirally) become established, and are very prejudicial to distinct vision. The remedy is obvious; viz., to dispense with a *tube* altogether, substituting for it a light, strong, inflexible framework of cast or wrought iron. In refracting telescopes, in which the air is completely enclosed, its circulation is not nearly so offensive.

5 I annex the readings of a thermometer on the night of July, 1834, on the desk, in the open air, in the gallery of the 20-feet reflector. Minimum before midnight 40°.4. At midnight, the hot wind set in, after a brief conflict, of the nature described in the text; the thermometer instantly rose to 50°. At

h	**m**		**o**
1	39 A.M.	it stood at	64.0
1	53 A.M.	it stood at	68.7
1	59 A.M.	it stood at	65.0
2	1 A.M.	it stood at	68.1
2	3 A.M.	it stood at	69.0
2	10 A.M.	it stood at	70.4
2	18 A.M.	it stood at	70.0
2	22 A.M.	it stood at	71.0
2	38 A.M.	it stood at	69.4
2	47 A.M.	it stood at	71.0
3	18 A.M.	it stood at	68.0

Minimum before sunrise next morning, 60.0

6 On such occasions, optical phenomena, of extraordinary splendour, are produced by intercepting the light of a bright star, by diaphragms pierced in regular patterns, and extending over the whole aperture of the telescope, such, for example, as large sheets of card-board or zinc, pierced by machinery (which gives perfect regularity), either with

circular holes, uniformly disposed, or with any regular and not too complicated pattern. The appearances so produced (which strike every one who witnesses them with surprise and delight), though they may be seen whenever the air is in a moderately good state, are infinitely enhanced in their beauty by the perfect tranquillity which prevails on such occasions as these. They depend on the optical law of interferences; and many beautiful examples of their explanation on that principle, fully worked out in detail, will be found in M. Schwerd's treatise, "Die Beugungserscheinungen aus den Fundamentalgesetzen der Undulations—theorie analytisch entwickelt," &c. (Manheim, 1835).

The "triangular aperture," or diaphragm which admits the light through an opening concentric with the speculum in the form of an equilateral triangle, to whose use as a means of separating close double stars continual reference will be found in the following pages, affords an elegant example of this theory, in the sharpness of the central disc which it produces, and the absence of all appendages other than six perfectly straight delicate rays running off at angles of 60° from the disc. In a letter addressed, Dec. 24, 1834, to the late Captain B. Hall, of which I retain a copy, I find an observation of Canopus, with such an aperture, and a magnifying power of 1200, thus described:—"The disc is an exact circle, and the six rays which such an aperture always gives, are perfectly straight, delicate, brilliant lines, like brightly illuminated threads, running far out beyond the field of view, and (what is singular) capable of being followed, like real appendages to the star, long after the star itself had left the field. In examining stars to see if they are close double, I always apply the triangular aperture. It reduces the discs to hardly more than a third of their size, and gives them a clearness and perfection incredible without trial."

Editorial Headnote

Agnes Mary Clerke, *A Popular History of Astronomy During the Nineteenth Century* (Edinburgh: Adam & Charles Black, 1885), 1–10

Agnes Mary Clerke, 1842–1907, was an Irish astronomer. She charts the development of institutional astronomy from gentlemanly enterprise, mirroring a shift from mathematical astronomy to physical astronomy.

24

A POPULAR HISTORY OF ASTRONOMY DURING THE NINETEENTH CENTURY

Agnes Mary Clerke

Source: Edinburgh: Adam & Charles Black, 1885, 1–10

We can distinguish three kinds of astronomy, each with a different origin and history, but all mutually dependent, and composing, in their fundamental unity, one science. First in order of time came the art of observing the returns and measuring the places of the heavenly bodies. This was the sole astronomy of the Chinese and Chaldeans; but to it the vigorous Greek mind added a highly complex geometrical plan of their movements, for which Copernicus substituted a more harmonious system, without as yet any idea of a compelling cause. The planets revolved in circles because it was their nature to do so, just as laudanum sets to sleep because it possesses a *virtus dormitiva.* This first and oldest branch is known as "observational," or "practical astronomy." Its business is to note facts as accurately as possible; and it is essentially unconcerned with schemes for connecting those facts in a manner satisfactory to the reason.

The second kind of astronomy was founded by Newton. Its nature is best indicated by the term "gravitational;" but it is also called "theoretical astronomy."[1] It is based on the idea of cause; and the whole of its elaborate structure is reared according to the dictates of a single law, simple in itself, but the tangled web of whose consequences can be unravelled only by the subtle agency of an elaborate calculus.

The third and last division of celestial science may properly be termed "physical and descriptive astronomy." It seeks to know what the heavenly bodies are in themselves, leaving the How? and the Wherefore? of their movements to be otherwise answered. Now such inquiries became possible only with the invention of the telescope, so that Galileo was, in point of fact, their originator. But Herschel was the first to give them a prominence which the whole progress of science during the nineteenth century has served to confirm and render more exclusive. Inquisitions begun with the telescope have been extended and made effective in unhoped-for directions by the aid of the spectroscope and photographic camera;

DOI: 10.4324/9781003009337-31

and a large part of our attention in the present volume will be occupied with the brilliant results thus achieved.

The unexpected development of this new physical-celestial science is the leading fact in recent astronomical history. It was out of the regular course of events. In the degree in which it has actually occurred it could certainly not have been foreseen. It was a seizing of the prize by a competitor who had hardly been thought qualified to enter the lists. Orthodox astronomers of the old school looked with a certain contempt upon observers who spent their nights in scrutinising the faces of the moon and planets, rather than in timing their transits, or devoted daylight energies, not to reductions and computations, but to counting and measuring spots on the sun. They were regarded as irregular practitioners, to be tolerated perhaps, but certainly not encouraged.

The advance of astronomy in the eighteenth century ran in general an even and logical course. The age succeeding Newton's had for its special task to demonstrate the universal validity, and trace the complex results of the law of gravitation. Its accomplishment occupied just one hundred years. It was virtually brought to a close when Laplace explained to the French Academy, November 19, 1787, the cause of the moon's accelerated motion. As a mere machine, the solar system, so far as it was then known, was found to be complete and intelligible in all its parts; and in the *Mécanique Céleste* its mechanical perfections were displayed under a form of majestic unity which fitly commemorated the successive triumphs of analytical genius over problems amongst the most arduous ever dealt with by the mind of man.

Theory, however, demands a practical test. All its data are derived from observation; and their insecurity becomes less tolerable as it advances nearer to perfection. Observation, on the other hand, is the pitiless critic of theory; it detects weak points, and provokes reforms which may be the beginnings of discovery. Thus, theory and observation mutually act and react, each alternately taking the lead in the endless race of improvement.

Now, while in France Lagrange and Laplace were bringing the gravitational theory of the solar system to completion, work of a very different kind, yet not less indispensable to the future welfare of astronomy, was being done in England. The Royal Observatory at Greenwich is one of the few useful institutions which date their origin from the reign of Charles II. The leading position which it still occupies in the science of celestial observation was, for near a century and a half after its foundation, an exclusive one. It was absolutely without a rival. Systematic observations of sun, moon, stars, and planets were, during the whole of the eighteenth century, made only at Greenwich. Here materials were accumulated for the secure correction of theory, and here refinements were introduced by which the exquisite accuracy of modern practice in astronomy was eventually attained.

The chief promoter of these improvements was James Bradley. Few men have possessed in an equal degree with him the power of seeing accurately, and reasoning on what they see. He let nothing pass. The slightest inconsistency between what appeared and what was to be expected, roused his keenest attention; and he never relaxed his mental grip of a subject until it had yielded to his persistent inquisition. It was to these qualities that he owed his discoveries of the aberration of light and the nutation of the earth's axis. The first was announced in 1729. It means that, owing to the circumstance of light not being instantaneously transmitted, the heavenly bodies appear shifted from their true places by an amount depending upon the ratio which the velocity of light bears to the speed of the earth in its orbit. Because light travels with enormous rapidity, the shifting is very slight; and each star returns to its original position at the end of a year.

Bradley's second great discovery was finally ascertained in 1748. Nutation is a real "nodding" of the terrestrial axis produced by the dragging of the moon at the terrestrial equatorial protuberance. From it results an *apparent* displacement of the stars, each of them describing a little ellipse about its true, or "mean" position, in a period of eighteen years and about seven months.

Now an acquaintance with the fact and the laws of each of these minute irregularities is vital to the progress of observational astronomy; for without it the places of the heavenly bodies could never be accurately known or compared. So that Bradley, by their detection, at once raised the science to a higher grade of precision. Nor was this the whole of his work. Appointed Astronomer-Royal in 1742, he executed during the years 1750–62 a series of observations which formed the real beginning of exact astronomy. Part of their superiority must, indeed, be attributed to the co-operation of John Bird, who provided Bradley in 1750 with a measuring instrument of till then unequalled excellence. For not only was the art of observing in the eighteenth century a peculiarly English art, but the means of observing were furnished almost exclusively by British artists. John Dollond, the son of a Spitalfields weaver, invented the achromatic lens in 1748, removing thereby the chief obstacle to the development of the powers of refracting telescopes; James Short, of Edinburgh, was without a rival in the construction of reflectors; the sectors, quadrants, and circles of Graham, Bird, Ramsden, and Cary were inimitable by Continental workmanship.

Thus practical and theoretical astronomy advanced on parallel lines in England and France respectively, the improvement of their several tools—the telescope and the quadrant on the one side, and the calculus on the other—keeping pace. The whole future of the science seemed to be theirs. The cessation of interest through a too speedy attainment of the perfection towards which each spurred the other, appeared to be the only danger it held in store for them. When, all at once, a rival stood by their side—not, indeed, menacing their progress, but threatening to absorb their popularity.

The rise of Herschel was the one conspicuous anomaly in the astronomical history of the eighteenth century. It proved decisive of the course of events in the nineteenth. It was unexplained by anything that had gone before; yet all that came after hinged upon it. It gave a new direction to effort; it lent a fresh impulse to thought. It opened a channel for the widespread public interest which was gathering towards astronomical subjects to flow in.

Much of this interest was due to the occurrence of events calculated to arrest the attention and excite the wonder of the uninitiated. The predicted return of Halley's comet in 1759 verified, after an unprecedented fashion, the computations of astronomers. It deprived such bodies for ever of their portentous character; it ranked them as denizens of the solar system. Again, the transits of Venus in 1761 and 1769 were the first occurrences of the kind since the awakening of science to their consequence. Imposing preparations, journeys to remote and hardly accessible regions, official expeditions, international communications, all for the purpose of observing them to the best advantage, brought their high significance vividly to the public consciousness; a result aided by the facile pen of Lalande, in rendering intelligible the means by which these elaborate arrangements were to issue in an accurate knowledge of the sun's distance. Lastly, Herschel's discovery of Uranus, March 13, 1781, had the surprising effect of utter novelty. Since the human race had become acquainted with the company of the planets, no addition had been made to their number. The event thus broke with immemorial traditions, and seemed to show astronony as still young, and full of unlooked-for possibilities.

Further popularity accrued to the science from the sequel of a career so strikingly opened. Herschel's huge telescopes, his detection by their means of two Saturnian and as many Uranian moons, his piercing scrutiny of the sun, picturesque theory of its constitution, and sagacious indication of the route pursued by it through space; his discovery of stellar revolving systems, his bold soundings of the universe, his grandiose ideas, and the elevated yet simple language in which they were conveyed—formed a combination powerfully effective to those least susceptible of new impressions. Nor was the evoked enthusiasm limited to the British Isles. In Germany, Schröter followed—*longo intervallo*—in Herschel's track. Von Zach set on foot from Gotha that general communication of ideas which gives life to a forward movement. Bode wrote much and well for unlearned readers. Lalande, by his popular lectures and treatises, helped to form an audience which Laplace himself did not disdain to address in the *Exposition du Système du Monde*.

This great accession of popularity gave the impulse to the extraordinarily rapid progress of astronomy in the nineteenth century. Official patronage combined with individual zeal sufficed for the elder branches of the science. A few well-endowed institutions could accumulate the materials needed by a few isolated thinkers for the construction of theories of wonderful beauty and elaboration, yet precluded, by their abstract nature, from winning general applause. But the new physical astronomy depends for its prosperity upon the favour of the multitude whom its striking results are well fitted to attract. It is, in a special manner, the science of amateurs. It welcomes the most unpretending co-operation. There is no one "with a true eye and a faithful hand" but can do good work in watching the heavens. And not unfrequently prizes of discovery which the most perfect appliances failed to grasp have fallen to the share of ignorant or ill-provided assiduity.

Observers, accordingly, have multiplied; observatories have been founded in all parts of the world; associations have been constituted for mutual help and counsel.

A formal astronomical congress met in 1798 at Gotha—then, under Duke Ernest II. and Von Zach, the focus of German astronomy—and instituted a combined search for the planet suspected to revolve undiscovered between the orbits of Mars and Jupiter. The Astronomical Society of London was established in 1820, and the similar German institution in 1863. Both have been highly influential in promoting the interests, local and general, of the science they are devoted to forward; while functions corresponding to theirs have been discharged elsewhere by older or less specially constituted bodies, and new ones are springing up on all sides.

Modern facilities of communication have helped to impress more deeply upon modern astronomy its associative character. The electric telegraph gives a certain ubiquity which is invaluable to an observer of the skies. With the help of a wire, a battery, and a code of signals, he sees whatever is visible from any portion of our globe, depending, however, upon other eyes than his own, and so entering as a unit into a widespread organisation of intelligence. The press, again, has been a potent agent of co-operation. It has mainly contributed to unite astronomers all over the world into a body animated by the single aim of collecting "particulars" in their special branch for what Bacon termed a History of Nature, eventually to be interpreted according to the sagacious insight of some one among them gifted above his fellows. The first really effective astronomical periodical was the *Monatliche Correspondenz*, started by Von Zach in the year 1800. It was followed in 1822 by the *Astronomische Nachrichten*, later by the *Memoirs* and *Monthly Notices* of the Astronomical Society, and by the host of varied publications which now, in every civilised country, communicate the discoveries made in astronomy to divers classes of readers, and so incalculably quicken the current of its onward flow.

Public favour brings in its train material resources. It is represented by individual enterprise, and finds expression in an ample liberality. The first regular observatory in the southern hemisphere was founded at Paramatta by Sir Thomas Makdougall Brisbane in 1821. The Royal Observatory at the Cape of Good Hope was completed in 1829. Similar establishments were set to work by the East India Company at Madras, Bombay, and St. Helena, during the first third of the nineteenth century. The organisation of astronomy in the United States of America was due to a strong wave of popular enthusiasm. In 1825 John Quincy Adams vainly urged upon Congress the foundation of a National Observatory; but in 1843 the lectures of Ormsby MacKnight Mitchel on celestial phenomena stirred an impressionable audience to the pitch of providing him with the means of erecting at Cincinnati the first astronomical establishment worthy the name in that great country. On the 1st of January 1882 no less than one hundred and forty-four were active within its boundaries.

The apparition of the great comet of 1843 gave an additional fillip to the movement. To the excitement caused by it the Cambridge Observatory—called the "American Pulkowa"—directly owed its origin; and the example was not ineffective elsewhere. Corporations, universities, municipalities, vied with each other in the creation of similar institutions; private subscriptions poured in; emissaries were sent to Europe to purchase instruments and procure instruction in their use.

In a few years the young Republic was, in point of astronomical efficiency, at least on a level with countries where the science had been fostered since the dawn of civilisation.

A vast widening of the scope of astronomy has accompanied, and in part occasioned, the great extension of its area of cultivation which our age has witnessed. In the last century its purview was a comparatively narrow one. Problems lying beyond the range of the solar system were almost unheeded, because they seemed inscrutable. Herschel first showed the sidereal universe as accessible to investigation, and thereby offered to science new worlds—majestic, manifold, "infinitely infinite" to our apprehension in number, variety, and extent—for future conquest. Their gradual appropriation has absorbed, and will long continue to absorb, the powers which it has served to develop.

But this is not the only direction in which astronomy has enlarged, or rather has levelled, its boundaries. The unification of the physical sciences is perhaps the greatest intellectual feat of recent times. The process has included astronomy; so that, like Bacon, she may now be said to have "taken all knowledge" (of that kind) "for her province." In return, she proffers potent aid for its increase. Every comet that approaches the sun is the scene of experiments in the electrical illumination of rarefied matter, performed on a huge scale for our benefit. The sun, stars, and nebulæ form so many celestial laboratories, where the nature and mutual relations of the chemical "elements" may be tried by more stringent tests than sublunary conditions afford. The laws of terrestrial magnetism can be completely investigated only with the aid of a concurrent study of the face of the sun. The positions of the planets will perhaps one day tell us something of impending droughts, famines, and cyclones.

Astronomy generalises the results of the other sciences. She exhibits the laws of Nature working over a wider area, and under more varied conditions, than ordinary experience presents. Ordinary experience, on the other hand, has become indispensable to her progress. She takes in at one view the indefinitely great and the indefinitely little. The mutual revolutions of the stellar multitude during tracts of time which seem to lengthen out to eternity as the mind attempts to traverse them, she does not admit to be beyond her ken; nor is she indifferent to the constitution of the minutest atom of matter that thrills the ether into light. How she entered upon this vastly expanded inheritance, and how, so far, she has dealt with it, is attempted to be set forth in the ensuing chapters.

Note

1 The denomination "physical astronomy," first used by Kepler, and long appropriated to this branch of the science, has of late been otherwise applied.

Editorial Headnote

Eliza Bowen, *Astronomy by Observation: An Elementary Text-Book for High-Schools and Academies* (New York: D. Appleton, 1888), 5–8

Eliza Bowen, dates unknown, was an American science educator and writer, active especially in the 1880s and 1890s. Here she extols the virtues of observation for those beginning an education in astronomy.

25

ASTRONOMY BY OBSERVATION

An Elementary Text-Book for High-Schools and Academies

Eliza Bowen

Source: New York: D. Appleton, 1888, 5–8

INTRODUCTION

For teachers and private learners

How to learn and teach the constellations

WE can not help seeing the motions of the heavenly bodies, unless, like Bunyan's man with the muck-rake, we steadily look down. A professed study of astronomy which does not make us recognize these bodies and their motions, when we see them, may not be useless, but it bears some marks of counterfeit knowledge. A mere acquaintance with the constellations is a very subordinate part of the knowledge here recommended, but it has, besides its own value, a good deal of importance as a prerequisite to the rest.

There are two disadvantages in the way of a teacher who wishes to make his pupils in astronomy really study nature. He can not be with them at night without inconvenience, and work outside of school-hours. This is the first difficulty. Another is, the impossibility of having the planets brought out for inspection when we want them, and also because, when they come, their motions are very slow.

After struggling a good deal with these disadvantages just mentioned, I found that by making young students learn the heavens as they do the maps of the United States or Europe, viz., by map-drawing, the whole of the first difficulty was obviated, and a large part of the second. They found the star-groups without trouble, for themselves. They were trained in observation, and they learned a great many of the motions incidentally. They became independent observers, and were so much interested in the practical study, that there was no doubt they would see and study for themselves whatever they could not see during the school-session.

DOI: 10.4324/9781003009337-32

No doubt a great many teachers can carry on this work quite as well as I can, or better; but, for the benefit of those who can not, I will describe it somewhat in detail.

The brighter stars in all the constellations, when joined by imaginary lines, form figures which are very easily recognized when once suggested. They are thus drawn in dotted lines on the maps attached to this book. If the student takes one figure at a time, and draws it repeatedly, until he gains the power to reproduce it on paper from memory, rapidly, without erasures, and with a fair degree of rough correctness, he can easily find it for himself if it is a conspicuous constellation, and he has some general idea in what part of the sky to look for it. And even if it is not a conspicuous constellation, he can easily find it, if it is adjacent to a constellation already known, and in a known direction from it. Thus, beginning with conspicuous constellations, and proceeding to adjacent ones, and then to others adjacent to the last, they are all easily learned.

But this plan of teaching by map-drawing is not a mere substitute for what is called "pointing out the constellations," and is not adopted and recommended only because the other course is inconvenient or impossible when teacher and class lodge in different houses. It is so much the easiest way, that when instructor and pupils are already together in the evening, and the latter wish to learn the star-groups, the teacher will do well to sit in the house (as I have often done), draw the figures on paper or a blackboard, and send students out to look for them. At first they feel helpless, and object; but a little trial soon convinces them how easy it is. If any find a difficulty, the others help them; and thus a teacher is spared additional work.

But mere drawing by the teacher is not sufficient, nor is that the plan recommended here. By the method described in the two preceding paragraphs, it is easy to have star-groups recognized at the time, but they will not be remembered, and must be taught again. Drawing by the pupil not only makes it easier to find them; they can hardly be fixed permanently in knowledge during the school-study, except by repeated drawing from memory. Whenever students have a few leisure moments, they should be called upon to cover the blackboard with these figures, or to reproduce them rapidly upon paper.

In order to secure repeated observation, it is best to have the groups learned one at a time; for, whenever the student goes out to find a new group, he is sure to look again at the old ones. The groups are both fixed in memory, and the habit of observation is strengthened. But the study of the motions gives far more interesting work than the mere identification of star-groups.

Any method of map-drawing already used in a school will answer for this purpose when carried out under the restrictions to be hereafter mentioned. The plan which I practice and recommend is called "drawing by dictation," now much practiced in the best schools. The teacher leads, standing at the blackboard, while pupils draw at desks in unruled exercise-books, which should not be used for any other purpose. When a teacher is wholly incapable of drawing even these simple figures, he can select some pupil to take the lead, who learns beforehand how to draw this group. Sufficient command over a pencil for this purpose is very common in all schools. Of course, the teacher is present to direct. The leader at the blackboard draws one of the lines of the figure and pauses a moment, while pupils follow with pencil and paper. Then he draws another line, and they follow, and so on until all the lines are finished. These lines are dotted on the map, but, in the exercise described, they are drawn continuous but very lightly. As the leader draws, he calls attention to the dimensions and positions of lines. No erasure whatever should be allowed. This is an important direction, for, unless it is observed, the teacher will be discouraged by the time consumed. It is not necessary to have more than a rough correctness, though, of course, pupils must not be encouraged to draw carelessly. After the lines are drawn, they must make the stars, the leader calling attention to the number and position. In the drawings made by leader and pupils, stars of the first magnitude are best represented by the symbol ○; of the second, by *; of the third, by ×; of the fourth, by •. In the "Description of Constellations," in the Appendix, there is given the number of first and second magnitude stars for each figure, and other information of value in studying them. After each constellation is drawn, the name should be written two or three times, and also the names of any first and second magnitude stars mentioned in the "Description." Then students should be called upon to repeat names aloud, separately and in unison. Without this exercise, they struggle a good deal with the names.

The drawing of a constellation must be made two or three times rapidly, and then the students are directed to turn over a leaf and draw from memory. The whole exercise, properly managed, need not take more than ten minutes. It should take place on the morning of a day which looks as if it would be followed by a clear night. If rain and interruption occur, the exercise should be repeated when again there seems a chance for observation at night. Each repetition takes less time. After drawing a figure, the student is always told to turn it in every direction, and is warned that he can not be sure beforehand exactly how it will be turned.

There is one important warning in regard to students who have identified one constellation and are beginning to learn an adjacent one. If a teacher, trying this method, fails in making his students learn the constellations, it will probably be from carelessness about this matter now to be explained. A picture of a constellation looked down upon, so reverses the directions of the same figure looked up to on the heavens, that the student who remembers the picture finds himself, as he says, "turned round" when he looks up. If he is merely studying a single figure, occupying a very small portion of the heavens, he can in a few moments recover his bearings, and recognize the group sought.

But, when the study by picture comes to be applied to a larger portion of the heavens, or more than a single figure, unless the student has the additional aid now to be explained, he will get not only "turned round," but seriously confused and perhaps discouraged as to his power to find anything at all without some one at his elbow. After learning to draw a group from memory, the student must make sure of the direction in which it lies from the known group. It should be the teacher's business to drill him on the direction of the new group from the known one. He should point out the side of the old one on which it is to be found, and he may draw the two together.

After he has made the drawing, and knows the direction of the new group from the old one, he should proceed rigorously in the following order when he studies the heavens: 1. He finds the known constellation. 2. Dismissing from his imagination all remembrance of figures, he finds carefully the proper direction from the known group. When he is sure of the direction, and not until then, he recalls the new figure which he has drawn, and he usually finds it in a few minutes.

The student should be well drilled in this order of proceeding, made to repeat and understand it.

The student should not *begin* by studying the maps of the book. He will think they are intended to aid him in finding the direction of groups *from himself* and the horizon. They do, it is true, show him the general aspect of the heavens at the times indicated on them. This is the aspect of the circles at the time of the equinoxes and solstices; and one object of the maps is, that the student may, after learning star-groups separately, know these. It is also the object of the maps to show the figures of the constellations and their directions from each other, and also the position of the various groups in regard to the ecliptic, the equinoctial, and colures, but not in regard to the observer and horizon.

When the first constellation is learned, the teacher (who looks it out beforehand) tells pupils where to look for it; and, whenever a constellation is studied, not adjacent to known ones, this must be repeated. But, after they learn a few, students cease to think of the position in regard to the inclosing horizon, and see that the maps can be used to learn new constellations adjacent to old ones. They then disregard the relation to the horizon, and often extend their knowledge in advance of the drawing. But the teacher should not stop the drawing. The securities for permanent and accurate knowledge are very insufficient without it.

The direction of star-groups from the observer and his horizon could not be permanently given by maps, for these directions are all the time changing. It is true, a globe can give them so that the change will not be noticed for a little while. But the globe can not be made a satisfactory help in studying the star-groups unless we use the directions from adjacent groups as well as directions from the horizon.

The dictation exercise in drawing, which has been recommended and described, is a species of free-hand drawing, which affords valuable training to hand and eye. Students can certainly learn the figures by drawing directly from the maps, but they will make little cramped figures, and find it difficult to draw any others, even on the blackboard.

I will state that, in using this book, I should have pupils study Chapter I, and then devote two weeks to study of constellations by map-drawing, before beginning Chapter II.[1] It is not easy to have good book-study done concurrently with drawing and identifying star-groups. The knowledge of the constellations is of indispensable importance in studying the motions of the solar system in nature, though it may easily be overestimated as an end in itself. When they are known, the student observes for himself the motions described in the text. Therefore it is best to have those groups that are visible learned in the beginning. After this it is not necessary to take more than a few moments from book-study, in order to call the student's attention to any phenomena in the heavens which it is desirable he should see.

Of course, as other constellations come into the evening sky in the east, they must be learned. The occasional suspension of book-study for one clear night will be sufficient in order to keep up with changes.

Where there is no time to have all the constellations studied in class, the polar and zodiacal star-groups, and those containing first and second magnitude stars, should be selected. As the constellations of the zodiac are not all conspicuous, their superior importance may be overlooked by an inexperienced teacher.[2] They should be learned as soon as possible.[3] When the student knows them separately, he should at once begin to draw them on the ecliptic. He should draw the line first, and then the star-groups on it in correct order and position. Students should often be required to draw the zodiac and ecliptic on strips of paper in a few moments. It is a good plan to have the line occasionally drawn on a long blackboard, and divided into twelve parts, and to send twelve students together to draw figures rapidly, giving them no warning as to which they are to draw, and timing them. But it is not desirable to draw the constellations on the equinoctial and colures, until the student has been very familiar with the ecliptic for a long time. The ecliptic is the important circle, and an imperfect knowledge of the equinoctial and colures would poorly compensate for a failure to be very familiar with the ecliptic. Therefore, the other circles should not be studied until there is no possible chance of getting confused with the ecliptic.

I hope I may be pardoned for telling of a test to which I put this method of teaching constellations. I took a class of fifteen children, whose average age was twelve years, but of whom three were only eleven years old, and I began to teach them the constellations at the beginning of the school year in September. Until a test was made at the beginning of November, I was never with them at night on a single occasion, nor had they any one at home who gave them help. They simply drew the figures and stars, and were drilled as here described. There was a good deal of bad weather, and another study took the place of star-learning, at least a third of the time. The lesson was half an hour long. At the first of November they knew perfectly every constellation which had been visible. They could tell where all the first and second magnitude stars were found. They knew nearly every third-magnitude star and a good many fourth. They could trace the ecliptic in the heavens, or draw it. They knew Algol, Mira, Epsilon Lyræ, the star in Draco which was the pole-star; Var in Aquila, midway between the poles; and what made them all remarkable. They knew the points where the sun is found, December 22d and March 21st, and their names. They found no difficulty whatever in learning Pisces and Aquarius. Some had a little trouble with Hercules and Ophiuchus, but finally found them. They noticed the daily and annual motion of the stars, the revolution of the moon and her motion among the stars, and they watched with great interest the motion of Venus among the stars. They also noticed that the constellations of the zodiac seemed to approach the sun in succession, and other points not necessary to mention. Their enthusiasm was very great, and I am sure they are observers for life.

This book is, of course, not intended for children, and the subject is only mentioned to show what can be done by the method here recommended of promoting observation.

In schools, where advancement depends on examination, observation-work will almost inevitably suffer, unless there is also an examination in it.

For the benefit of inexperienced teachers, it is well to give some advice in regard to the constellations which it is best to begin with. Of course, the Great Dipper, the Little Dipper, and Cassiopeia should come first. Chapter II contains directions how to learn them. After these are learned, some conspicuous constellation must be selected and found, and the rest of the study proceeds from that by adjacent constellations. For the three autumn months, Pegasus is best; for the winter months, Orion; for spring, Leo Major; for summer, Scorpio. By finding these in the "Description of Constellations alphabetically arranged," Appendix A, the learner and inexperienced teacher will find hints how to proceed.

Most teachers without experience in observation-teaching will, when using this book for the first time, secure only part of the observation-study. The book can be used for a mere cram-book, and I have endeavored to make it stimulate the student even in that case, and give him power and zeal to help himself. Every teacher should try to secure some beginning of observation, and with each successive class the work will increase.

A school year devoted to astronomy will give very admirable results; but with five months, work can be done well worth undertaking. It will be good work if a class sees enough of the daily motion of stars to perceive that it is a revolution with the pole for a center; enough of the annual motion to see that stars watched at intervals of two or three weeks at the same hour move west, disappear beyond the western horizon, and reappear above the eastern; some motion of the sun on the horizon; something of the increasing or decreasing obliquity of the sun's rays. The moon comes so conveniently for observation that, where it is possible, I recommend that she be studied through an entire lunar period from night to night. This work trains in observation, and the knowledge in nature, of the important positions, opposition and conjunction, is useful when the pupil studies the planets.

This observation-study can be directed without going out with students at night at all.

No elementary science is so independent of expensive apparatus in schools as astronomy. The Copernican theory is the groundwork of all formal treatises on astronomy, and its author died before the invention of the telescope. No telescope, used in ordinary schools could aid in its rational acceptance, except by showing the phases of Venus and the moons of Jupiter. The bearing of these on the theory could not be at all appreciated by a person who had not previously studied what can be seen with the naked eye. It is, of course, a pleasure and advantage, after learning what can be seen without a telescope, to see these objects, and also the rings of Saturn, the surface of the moon, and the resolution of some nebulæ; but I doubt whether it is any advantage at all until the student has used his unaided eyes. The most important need for high-schools seems to me to be some simple means of taking angular measurements on the heavens. This brings out a little the relation of the study to mathematics.

Of course, a large part of the information in any elementary astronomy comes from investigators equipped with the refined and complicated instruments of modern observation. In this case ordinary people can not see for themselves, but must take the facts on testimony. But here the phenomena seen can be so well represented by picture that ordinary people are under no great disadvantage in following with intelligence the conclusions of astronomers. Of course, we could neither intelligently accept the evidence nor appreciate the conclusions if we are entirely ignorant of the principles on which the instruments are constructed; but these can be easily understood by the aid of a few inexpensive mirrors, lenses, and prisms.

It is my own experience that young persons have a sort of superstition about apparatus which looks complicated, and especially as if it cost a good deal of money. The feeling is diametrically opposed to the spirit of investigation, and therefore until they have become possessed of that spirit it is best to teach them with the simplest appliances.

Of illustrative apparatus, a globe, with the ecliptic traced on it in connection with the equinoctial system of circles, is important; but there is a very small, cheap terrestrial globe now much used in primary schools which has the ecliptic traced on it for the purpose of using the globe to show the circles on the heavens. It can be made to answer the purpose, using an India-rubber band for the horizon when we wish to show the angles it makes with the ecliptic. The celestial horizon in nature is such a well-defined circle, and the sphere so plainly revolves through it, that there is not much lost by the globe not revolving through the India-rubber band. Of course, a good celestial globe is a great advantage, provided it does not take the place of nature-study. Where we can see for ourselves, illustrative apparatus is useful only to give definiteness to our ideas derived from observation; but the study of nature should come first. I have many times seen expensive illustrative apparatus used so as to discourage observation. This was not intended, but practically this was the result.

N. B.– The various figures on the maps sometimes occur on more than one map. For the student to copy, it is best to take them from maps where they are found on or near the middle of the map.

Notes

1 This plan would be modified by bad weather, making observation impossible.

2 The minuteness of these directions is intended for this class, whom the author hopes to aid. It is no reflection upon others.

3 The facts which show that the earth and planets revolve round the sun can be seen by anybody with eyes. The reasoning which establishes this theory, and the connection of these facts with it, are so simple and plain that any high-school student learning geometry

can understand them clearly. Thus he can look at the changes in the heavens with intelligence. But this whole result depends on knowing the constellations of the zodiac.

In using Charts I, II, III, IV, it is best to forewarn students that, where figures must be represented very small, as on these charts, the stars appear relatively somewhat larger than in nature. A teacher will always find it wiser for students to draw on a larger scale. The figures on the chart facing page 10 are on a very good scale for their drawings.

The fuller charts at the end of the book are for reference after the student has learned the brighter stars in the constellations, by means of Charts I, II, III, IV. The dotted lines show the boundaries of constellations, the Greek names of stars are given, and there is a fuller representation of the stars. The student's attention should be called to the fact that these charts represent the northern and southern hemispheres, the north pole being at the zenith.

Part 6

ANTHROPOMETRY/EUGENICS

Editorial Headnote

British Association for the Advancement of Science *Final Report of the Anthropometric Committee* (London: Spottiswoode, 1883), 1–3

The British Association for the Advancement of Science was founded in 1831 and retained this name until 2009. The Committee, comprising such luminaries as Francis Galton, was in pursuit of a statistical map of the 'races' of the United Kingdom, on the understanding central to anthropometry that physical measurements revealed hereditary norms and deficiencies. The endeavour represents the institutionalisation of statistical collection packaged with an implicit race science. Ideology, in many ways, precedes observation, but the presentation of facts is given the other way around, as if observation provides irrefutable proof of the physical signs of degeneracy. It is an important foundational document in what later become the science of eugenics.

26

BRITISH ASSOCIATION FOR THE ADVANCEMENT OF SCIENCE FINAL REPORT OF THE ANTHROPOMETRIC COMMITTEE

Source: London: Spottiswoode, 1883, 1–3

Final Report of the Anthropometric Committee, consisting in 1882–3 *of* Mr. F. Galton (*Chairman*), Dr. Beddoe, Mr. Brabrook (*Secretary*), Mr. Frank Fellows, Mr. James Heywood, Professor Leone Levi, Dr. F. A. Mahomed, Mr. J. E. Price, Lieut.-General Pitt-Rivers, Sir Rawson W. Rawson, *and* Mr. C. Roberts. *Associates*, Dr. T. G. Balfour, Dr. J. H. Gladstone, Inspector-General Lawson, Dr. W. Ogle.[1] *Drawn up by* Mr. C. Roberts *and* Sir Rawson W. Rawson.

1. The Committee, originally appointed in 1875, and aided by successive grants, of which it has expended 280*l*., has made a Report in each of the five years 1878 to 1882, and now submits its final Report.

2. Not that the work open to the Committee is exhausted, although it has to a great extent supplied what was pointed out in its Reports of 1881 and 1882 as chiefly wanting, or that its conclusions are to its own mind complete and satisfactory. But it would require more time and larger funds than are at the disposal of the Committee to prosecute its inquiries, even with the materials now in its possession, to the end which it has had in view; and the Committee is of opinion that the most useful course will be to bring before the Association the results of its past labours, indicating at the same time the conclusions which it considers to be sufficiently established by the facts ascertained, and the deficiencies, both of data and methods, which remain to be supplemented, either by individual exertion, or by the reappointment of a similar Committee at some future period under the auspices of the Association.

3. In order to furnish a complete review of the information obtained, it will be necessary to refer to tables and data contained in previous reports. A list of these Reports is furnished in a note.[2]

OBJECTS AND OPERATIONS OF THE COMMITTEE

4. The Committee was appointed for the purpose of collecting observations on the systematic examination of the height, weight, and other physical characters of the inhabitants of the British Isles.

DOI: 10.4324/9781003009337-34

5. Its operations in each year are described in the introduction to its Report of 1881. The description and amount of the statistics which it has collected, and the names of the persons to whom it is indebted for the collection, are detailed chiefly at the commencement of its several Reports from 1880 to 1882.

6. Among the objects early aimed at by the Committee, and prosecuted by it up to the year 1881, was the collection and comparison of photographs of the typical races of the United Kingdom; but at the meeting of that year this inquiry was assigned to a separate Committee, upon whom will devolve the duty of reporting upon this branch of the general subject.

7. The points to which the Committee has addressed its inquiries are—

(1) Stature.
(2) Weight.
(3) Girth of chest.
(4) Colour of eyes } Complexion.
(5) Colour of hair }
(6) Breathing capacity.
(7) Strength of arm.
(8) Sight.
(9) Span of arms.

To these might have been added others, especially—

(10) Size and shape of head.
(11) Length of lower limbs as shown by the difference between the sitting and standing positions.
(12) Girth, length, and breadth of other parts of the body.

But the Committee was afraid of seeking to obtain more information than their contributors would be likely to furnish; and experience has shown that many of them have been unable to supply more than a portion of that which was requested. Few have furnished complete returns on all the subjects, but where one has failed another has succeeded, and sufficient data have been collected to give trustworthy statistical results on all the subjects of inquiry except those of breathing capacity and sight. An abstract of one of the complete returns will be given in its proper place, as exhibiting a good epitome of what the Committee has sought to obtain in all cases. (See Table XXIII.)

8. The large body of observations on stature, weight, and complexion collected by Dr. Beddoe, and those on stature, weight, and chest-girth collected by Mr. Roberts, previously to the formation of the Committee, have been made use of; and the Committee has thus had observations made on a total number of about 53,000 individuals of both sexes and of all ages, from which to construct their tables and to base their conclusions.

9. The statistics are unique in range and numbers, and have been obtained from a very large number of independent observers living in different parts of the country, without prejudice, and often in ignorance of the use which would be made of them; and they have been analysed and tabulated in a perfectly impartial manner, irrespective of all preconceived opinions. The Committee does not claim for them exemption from the liability to that amount of imperfection and probable error which must attach to all conclusions drawn from a disproportionate, and from a comparatively small number of observations. But great care has been taken in the examination and classification of all the returns to eliminate obvious errors, and to call attention in the body of the Report to any apparent discrepancies from faulty observation or deficient numbers.[3]

METHODS

10. The forms and instruments used have been explained in the Reports for 1878 and 1880; but practical difficulties have been found to exist in obtaining trustworthy observations with regard to breathing capacity. Experience has also led the Committee to believe that the use of Snellen's test-types for sight, Nos. 1 and 10, is more convenient, and will yield more trustworthy results, than that of the army test-dots, which were adopted in its original circulars.[4] Since 1879, also, the Committee has introduced the use of cards for recording the observations relating to single persons, which has been extensively adopted in Germany and the United States, and recently by the Investigation Committee of the British Medical Association, and which offers great facilities in analysing and grouping the facts observed. The Committee appends copies of the forms of the cards and of the methods of measurement and observation which they have employed. (See Appendix A.)

11. The difference between the *average* and *mean* of a number of observations, and its importance in dealing with the subjects under consideration, has been pointed out and discussed by Mr. Roberts in the Report for 1881, at p. 233;[5] and the special sense in which Mr. Roberts employs the term *mean*, being that value in an arithmetic series of observed values of which the observations are the most frequent, has been adopted by the Committee.[6]

12. In connection with the question of the applicability of the exponential law of error to statistical results relating to anthropometry, Mr. Francis Galton has contributed a valuable series of tables, with remarks, on the range in height, weight, and strength, in which he introduces his method of the calculation of deciles, quartiles, and medians.[7]

Notes

1 The late Dr. William Farr was a member, and Chairman of the Committee from 1875 to 1879.

2 1, Report for 1878, 5 pp. (numbered pp. 182–6 in the Annual Report of the Association). 2, Report, 1879, 35 pp.; *ibid.* pp. 175–209. 3, Report, 1880, 41 pp.; *ibid.* pp. 120–59. 4, Report, 1881, 48 pp.; *ibid.* pp. 225–72. 5, Report, 1882, 3 pp.; *ibid.* pp. 278–80. An Index to the Tables is given in Appendix C.

3 'If an exceedingly large number of measurements, weights, &c. be taken—supposing no bias, or any cause of error acting preferably in any one direction to exist—not only will the number of small errors vastly exceed that of large ones, but the results will be found to group themselves about the mean of the whole always according to one invariable law of numbers, and *that* the more precisely, the greater the total number of determinations . . . . . Rude and unskilful measurements of any kind, accumulated in very great numbers, are competent to afford precise mean results. The only conditions are the continual *animus mensurandi,* the absence of bias, the correctness of the scale with which the measures are compared, and the assurance that we have the entire range of error, at least in one direction, within the record.'—Sir J. F. W. Herschel, *Edin. Rev.* vol. xcii.

4 See the Report for 1881 for a discussion of this subject by Mr. Lawson and Mr. Roberts.

5 Also in a note at p. 121 of the Report for 1880.

6 Mr. Roberts has followed Quetelet in the use of the word *mean,* and its difference from an *average* is thus explained by Sir John Herschel. Speaking of Quetelet's *homme moyen* he says:—'Now, this result, be it observed, is a *mean* as distinguished from an *average*. The distinction is one of much importance, and is very properly insisted on by M. Quetelet, who proposes to use the word mean only for the former, and to speak of the latter (average) as the "arithmetical mean." . . . . An average may exist of the most different objects, as of the height of houses in a town, or the size of books in a library. It may be convenient to convey a general notion of the things averaged, but involves no conception of a natural and recognised central magnitude, all differences from which ought to be regarded as deviations from a standard. The notion of a mean, on the other hand, does imply such a conception, standing distinguished from an average by this very feature, viz., *the regular march of the groups, increasing to a maximum and then again diminishing*. An average gives us no assurance that the future will be like the past. A mean may be reckoned on with the most implicit confidence. All the philosophical value of statistical results depends on a due appreciation of this distinction, and acceptance of its consequences.'—*Edin. Rev.* vol. xcii. Mr. Galton, however, desires to state that considering many statistical groups which are regular in their distribution are at the same time normally asymmetrical, he does not recognise the expressions of 'mean value' and 'the value most likely to be observed' as strictly equivalent.

7 Report for 1881, p. 245.

Editorial Headnote

Alexander Graham Bell, 'Sheep-Breeding Experiments on Beinn Bhreagh', *Science*, 36 (1912), 378–384

Alexander Graham Bell, 1847–1922, is known principally as the pioneer of the telephone and the founder of the American Telephone and Telegraph Company in 1885. He is claimed as a feted scientist by Scotland, Britain, the United States and Canada. Experimentation was at the core of his work, and here Bell is consumed by the selective breeding of sheep, with particular emphasis on the number of nipples. The piece has obvious eugenic overtones, though is presented in the idiom of curiosity, observation and experiment. Compare 2.24 for Bell's explicit advocation of positive eugenics.

27

'SHEEP-BREEDING EXPERIMENTS ON BEINN BHREAGH'

Alexander Graham Bell

Source: *Science*, 36 (1912), 378–384

SHEEP-BREEDING EXPERIMENTS ON BEINN BHREAGH[1]

Introductory Remarks.—It is astonishing how ignorant we all are about common things. Just test the matter on yourself. Sheep are quite common; and we are all more or less familiar with their appearance, and should therefore be able to answer some questions about them. Well then—*How many front teeth has a sheep got in its upper jaw?*

You never counted them? You have not observed? Next time you come across a sheep just look and see, and you will find that she has *none at all!*—the upper gum is bare.

We are all familiar with the fact that a sheep suckles her young; and know therefore that she possesses nipples that yield milk. How many nipples has she, and where are they located?

Human beings, of course, have only two, located on the breast. Dogs and cats and other mammals that have a litter at birth have many nipples, located in pairs all along the belly. Cows have at least four, located on the belly between the hind legs. Where are the sheep's nipples placed, and how many are there?

I must confess that I was myself unable to answer these questions, until, in the year 1890, I made a personal examination of the sheep on Beinn Bhreagh.

It then became obvious that sheep, like human beings, have only two nipples; and that they are located, as in the case of the cow, on the belly between the hind legs.

It was also found at this examination in 1890 that some sheep have four nipples instead of two. Two of these were in the usual place and of the usual size; the extra pair lay in front, upon the belly; and the nipples were extremely small and undeveloped, more nearly resembling pimples upon the skin than nipples. They were embryonic in character and yielded no milk.

At once interesting questions began to arise: Could we, by mating four-nippled ewes with four-nippled rams, and by selecting from their progeny for breeding

DOI: 10.4324/9781003009337-35

purposes the lambs in which the extra nipples were most fully developed, gradually create a variety of sheep that would have four nipples of equal size, all yielding milk?

If we could, by selection, create a four-nippled variety of sheep, why not a six-nippled, an eight-nippled, or a multi-nippled variety?

It was also found in 1890 that the proportion of sheep having four nipples was larger among the twin-bearing than the single-bearing ewes; and this at once raised the question as to whether there was any correlation between the number of nipples and the number of lambs at a birth.

If the four-nippled variety should turn out to be twin-bearing, as a rule, would the six-nippled and eight-nippled varieties give us triplets and quadruplets; and would the multi-nippled ewes have a litter at birth?

Twins, etc., are usually smaller at birth than single lambs, even after they have become mature, probably because their mothers are unable to supply sufficient milk for two or more lambs when the ewes have only two functional mammæ.

Would a sheep supply more milk from four functional nipples than two; and if so, would she be able to care for two lambs as easily as one?

A twin-bearing stock, able to rear twins successfully, would undoubtedly be of great value in a country like Nova Scotia, where the winters are long and the cost of sheep-breeding correspondingly great. If the farmers could raise two lambs instead of one for every ewe wintered, sheep-breeding in Nova Scotia might become a profitable industry of great importance.

These considerations led to the sheep-breeding experiments upon Beinn Bhreagh. I must here express my indebtedness to Mr. J. G. Davidson, Superintendent of Beinn Bhreagh Nursery, who has, for the last ten years, had charge of the experiments under my direction.

A Four-nippled Variety of Sheep.—By pursuing the plan of selection outlined above, we succeeded in producing upon Beinn Bhreagh, in a very few years, a four-nippled variety of sheep in which the ewes had, as a normal condition, four nipples of nearly equal size, all yielding milk.

So few cases of reversion to the ordinary two-nippled type appeared among the lambs that I felt justified in bringing the matter to the attention of the National Academy of Sciences, at their meeting in Washington, D. C., April 21, 1904.[2] At the same time I presented to the National Academy of Sciences, in pamphlet form, the "Sheep Catalogue of Beinn Bhreagh, Victoria County, Nova Scotia, showing the origin of the Multi-nippled Sheep of Beinn Bhreagh, and giving all the descendants down to 1903."[3]

This four-nippled breed was not slowly evolved by the persistent mating together of sheep having extra nipples of embryonic character. *It sprang suddenly into existence;* for it was soon discovered that it was possible to pick up here and there, from the farmers of Cape Breton Island, ewes with four functional mammæ already fully developed. These were added to our flock and hastened the accomplishment of the result.

In a few years after the experiments were begun in 1890, we had so many four-nippled sheep that we were able to cut down the flock severely. We killed or sold all sheep with extra nipples in an embryonic or undeveloped form, and limited ourselves to ewes with four *functional* nipples alone.

After this point had been reached there were very few cases of reversion to the two-nippled type.

The flock speedily increased in numbers, and when, in process of time, it became largely composed of four-nippled sheep whose parents had also been four-nippled, the cases of reversion practically ceased and the breed was established.

At first it appeared that the four-nippled ewes were less fertile than ordinary sheep, for they had a smaller proportion of twins; but this turned out to be due to the fact that the process of selection had necessarily resulted at first in a flock composed mainly of young ewes, and young sheep rarely have twins. After the four-nippled ewes had grown to full maturity they were found to be as fertile in this respect as ordinary sheep, if not more so.

Most of the twins that were born on Beinn Breagh were small at birth. In the autumn, however, they were found, upon the average, to be fully equal in size and weight to the single lambs of the flock, thus demonstrating the important point that a breed of sheep had been produced *which could successfully rear twins*.

During the process of the establishment of the four-nippled breed the number of two-nippled and three-nippled lambs born in the flock gradually decreased; and five-nippled lambs took their place in increasing numbers. Then six-nippled lambs were produced, followed by the occasional appearance of seven-nippled and even eight-nippled lambs, indicating the possibility of producing breeds of sheep with a greater number of nipples than four, if desired.

Since the year 1890, the nipples of several thousand sheep on the island of Cape Breton have been examined, with the discovery that three-nippled, four-nippled and even five-nippled sheep are by no means uncommon. Six-nippled sheep, on the other hand, are extremely rare, only two having been discovered in twenty-two years which were not connected with our flock. We have never come across a seven-nippled sheep that was not descended from Beinn Bhreagh stock; and eight-nippled sheep seem to be quite unknown at present outside of Beinn Bhreagh.

The fact that four-nippled sheep, like black sheep, are to be found in every flock of considerable size, led me to push the selection in the Beinn Bhreagh flock towards the formation of a six-nippled variety, so as to secure a breed that could not be easily duplicated elsewhere.

Origin of the Six-nippled Variety of Sheep. (*Ewe No. 76*.)—As early as 1891 we discovered in the flock of a farmer a six-nippled ewe, with the four extra nipples very poorly developed. We purchased her and added her to our flock as "No. 76," as she was a yearling at the time.

She remained on Beinn Bhreagh for several years and gave us nine lambs in all before she died. She was mated with our best four-nippled rams, but never gave us a six-nippled lamb. The first six-nippled lamb born on Beinn Bhreagh, however (No. 610), was her direct descendant (granddaughter) without any admixture of other six-nippled blood; and the ewe, No. 610, gave us a six-nippled lamb when she was only a year old herself.

No. 76 was a white ewe, and in 1893 she was mated with a white ram, but the lamb she produced in 1894 was black. He turned out to be a ram with four nipples (No. 417), and he is largely responsible for the black blood that afterwards appeared in the Beinn Bhreagh flock.

(*Ewe No. 256.*)—In the course of that same year (1894) we heard of another six-nippled ewe which had been discovered in the flock of a farmer near North River, St. Ann's; but she was so wild that the people on the farm were unable to catch her for us.

As we desired to secure her before the breeding season arrived, we sent a man to the farm two or three times to assist in her capture, but all without success. She was as wild as a deer and leapt the fences and escaped to the woods.

Finally a large expedition was sent about October, 1894. The hunt occupied the greater part of a day, and at last the men succeeded in driving the ewe into a place from which there was no escape, and she was brought to Beinn Bhreagh and added to our flock as No. 256.

She turned out to be a black ewe, two years old, with six well-developed and functional nipples, well arranged in pairs. She was mated with the black ram No. 417 (the offspring of the other six-nippled ewe No. 76), so that any lamb she might have would combine the blood of the two six-nippled ewes, No. 76 and No. 256.

Of course the results of the union were eagerly looked for, but in the spring of 1895 the ewe, No. 256, escaped and lambed in the woods.

Then there was a hunt to save the possibly six-nippled lamb from the foxes that had occasionally taken toll of our flock. It took quite a large number of men, in skirmishing order, to re-capture the run-away, but the lamb turned out, after all, to be a black ewe *with only four nipples!*

Although the black six-nippled ewe (No. 256) lived for many years on Beinn Bhreagh, and gave us twelve lambs in all (including, by-the-bye, five sets of twins) we never got a six-nippled lamb from her—unless indeed her lamb No. 940 might be so considered.

This lamb (No. 940) was at first noted as a five-nippled ewe; but one of her nipples, instead of being round like the others, was greatly elongated in cross section, and had two distinct orifices. It was evidently formed by the union of two distinct nipples into one. After lambing, both of the Siamese-twin nipples were found to yield milk; and we have the ewe now recorded on our books as a six-nippled sheep.

In the autumn of 1895 the black six-nippled ewe No. 256 was again mated with the black four-nippled ram, No. 417; and in the spring of 1896 gave us black twins; one, a female with four nipples, and the other a male, No. 626, with five nipples.

This five nippled ram, No. 626, not only represented an advance in nipples over rams formerly employed; but, in addition, he combined in his own person the blood of the two six-nippled ewes, No. 76 and No. 256. He was, therefore, although black, used very extensively with the flock until white six-nippled rams appeared among his offspring, when they were substituted as the sires of the flock.

No. 810 and No. 827 (born 1898) were the first six-nippled rams employed in the flock; and six-nippled rams have been used ever since. In 1899, 25.6 per cent. of the lambs born were six-nippled, but in 1900 the percentage, for some unaccountable reason, fell to 4.4 per cent.; and the percentages in succeeding years, 1901, 1902 and 1903, were only 9.4 per cent., 9.6 per cent. and 11.1 per cent., showing a very slow rate of increase in spite of the fact that six-nippled rams had been used exclusively in the flock since the autumn of 1898.

There had been no difficulty in producing the four-nippled variety of sheep, because we had been able to obtain from surrounding farmers sheep with four nipples already fully developed to add to the flock, but in the case of the six-nippled variety we were unable to obtain this aid from the farmers.

No six-nippled sheep were to be had for love or money. We advertised for them and offered large prices. We notified butchers to examine the nipples of the sheep that came to them for slaughter, etc., but all in vain. During a period of thirteen years from the purchase of ewe No. 256, the only six-nippled sheep we were able to buy was a black ewe, No. 735, with very poorly developed extra nipples, which had been found by a butcher in Baddeck among the sheep purchased from farms in close proximity to Beinn Bhreagh, and which probably represented a leakage from our flock.

It will thus be seen that, as we were unable to obtain six-nippled blood from outside, the only way we could advance the formation of a six-nippled variety of sheep seemed to be: (1) To use six-nippled rams on the whole flock; (2) to select from the six-nippled ewes born on the place those that had the most fully developed extra nipples, discarding the others.

The second plan seemed to be impracticable on account of the small number of six-nippled ewes that made their appearance. We had perforce to keep all the six-nippled ewe lambs that appeared in order to preserve the six-nippled strain from the danger of extinction. No selection by six-nippled ewes was possible; and the bulk of the flock remained four-nippled for years in spite of the continual use of six-nippled rams.

Females More Important than Males.—We recognized the fact that we were breeding for a female peculiarity; and that in this case selection by females was probably more important than selection by males.

In the hope of increasing the number of lambs from which selection could be made we determined to enlarge our flock; and, in the autumn of 1903, we purchased several four-nippled sheep and a large number of ordinary two-nippled ewes, and used our six-nippled rams with the whole flock. This plan increased the number of lambs born in 1904 and reduced the percentage of six-nippled sheep to 4.8 per cent.

After one winter's experience it became obvious that it was inadvisable to have a large flock on account of the increased expense and the difficulty of preserving accurate records with large numbers.

In the autumn of 1904, therefore, we cut down the flock to one half; retaining only ewes having four or more functional nipples, and in the spring of 1905 the percentage of six-nippled lambs rose to 25.8 per cent.; followed by 23.6 per cent. in 1906, and 27.7 per cent. in 1907.

It was speedily realized that with a small flock continuous inbreeding was inevitable; and that it would be very advisable to introduce new blood, as the flock was undoubtedly deteriorating physically. The difficulty, however, was that no six-nippled sheep could be found outside our own flock. We searched the country far and wide, and offered fancy prices without any success.

Avoiding the Evils of Inbreeding.—In the autumn of 1906 we tried the experiment of loaning a six-nippled ram to a neighboring farmer; and we offered him $10.00 apiece for any six-nippled lambs produced, $15.00 for seven-nippled and $20.00 for eight-nippled lambs. As the result of this experiment we were able, in 1907, to purchase two six-nippled lambs.

The experiment of loaning a ram was tried again in the autumn of 1907, with the result that in 1908 we were able to purchase four six-nippled lambs, for which we paid the farmer $40.00. This excited the interest of other farmers, who now began to apply for the loan of our rams under similar conditions of payment for lambs.

This plan of loaning rams turned out to be a success; and by the autumn of 1909 it became obvious that, by pursuing this plan, we could rely upon surrounding farmers for a continual supply of new six-nippled blood without keeping a large flock ourselves.

Giving up the Four-nippled Breed.—In the autumn of 1909, therefore, we cut down our flock *to six-nippled ewes alone*, and distributed our four-nippled and five-nippled ewes, as a *bonus*, among the farmers who would use our rams.

In the spring of 1910, 50 per cent. of the lambs born on Beinn Bhreagh were six-nippled; and we were also able to purchase a number of six-nippled lambs from the farmers. Very similar results were obtained in 1911.

This spring (1912) 53.6 per cent. of our lambs were six-nippled; and the records handed in by the farmers who have used our rams indicate that we shall probably be able to purchase from fifteen to twenty six-nippled lambs this year.

The New Departure.—As it has now been demonstrated that we can rely upon a constant influx of new six-nippled blood from outside, the time has come when we may begin the selection of six-nippled sheep by females. We propose to cut down our small flock still more this year, and keep only six-nippled ewes that have all six nipples *in a functional condition.* We fully expect an increase in the percentage of six-nippled lambs born on Beinn Bhreagh next year, in spite of the fact of the large amount of alien blood in the flock. The new blood introduced has not diminished the proportion of six-nippled lambs born on the place, while it *has* resulted in increased vigor and strength in the flock as a whole. The physique has been improved to such an extent that there are now no better nor finer sheep than those to be found on Beinn Bhreagh.

A Discovery Relating to Color.—During the course of our breeding experiments a very curious fact made its appearance. It is well known that when white ewes are mated with white rams, black lambs occasionally appear among the offspring; but it has not hitherto been known that when black ewes are mated with black rams, the *offspring are always black.*

This fact was discovered by Dr. Davenport, Director of the Carnegie Institution's Station for Experimental Evolution at Long Island, N. Y., from a study of the records of the Beinn Bhreagh flock; and his conclusions have since been amply verified at Beinn Bhreagh and elsewhere.

Production of Twins.—The experience of the past has shown that twin lambs are usually the produce of old ewes; and the fact that 36 per cent. of the lambs born on Beinn Bhreagh are twins is very encouraging when we consider that they are the produce of young ewes.

We are getting twins from mothers that were only one year old, and two years old, at the time of the birth of their lambs; and 60 per cent. of the lambs born this spring from three-year-old ewes were twins.

We have only one old sheep on the place, No. 1505. She was seven years old this spring and has given us eleven lambs to date. She started out with twins when she was only a lamb herself (one year old) and has given us three sets of twins since then, besides three single lambs.

The indications are that our six-nippled stock will ultimately turn out to be twin bearers, as a rule, when they become fully mature.

In the meantime the question is: What can we do to favor the production of twins in our flock?

1. One point of importance will be to see that twin ewes are mated with twin rams.

2. Twin-bearing ewes are, on the average, much heavier than single-bearing ewes. We would therefore favor the production of twins by weeding out from the flock, while still young, ewes that are markedly under the average weight of the ewes of their age.

3. When our flock of adult ewes is large enough to permit of selection it might be well to discard ewes at the age of three years, at all events at four, if they have never given us twins; so that the bulk of our fully mature females should ultimately be composed of ewes that have produced twins when young.

These points relate to efforts to increase the *hereditary tendency* to the production of twins; but there are other points relating to *environment* which are also of importance.

Condition of Nutrition Important.—The records of weight that have been preserved at Beinn Bhreagh seem to indicate that the condition of nutrition of the mother at the time of mating has an important influence upon the conception of twins. Our records show that the twin-bearing ewes increase in weight as the mating period approaches; and that mating occurs when the ewes are in prime physical condition.

This probably explains the curious fact that the ewes, which are mated in October, have a much larger proportion of twin lambs than those which are mated later on in the breeding season; thus verifying the old saying of the farmers here that "March lambs bring twins." The pasture is at its best in October. It begins to fail in November, and by December there is very little left. Our records show that when ewes are mated in October the proportion of twins born is larger than when mating occurs in November, and that very few of the ewes mated in December have twins.

In order to favor the production of twins it is therefore necessary, or at least advisable, that the ewes should be in prime physical condition at the time of mating.

To secure this point we breed in October; and it has also been our custom for some years past to provide the ewes with extra nourishment in the form of oats, oil-cake, etc., for two or three weeks before mating.

In order to be perfectly sure that each ewe receives her proper share, individual feeding pens have been provided.

When a common trough is used the larger and stronger sheep, who need it least, get most of the food; and the smaller and weaker, who need it most, get least.

Nutrition After Mating.—Our records of weight indicate that there is a characteristic difference in the weights of twin-bearing and single-bearing ewes after mating.

The twin-bearing ewes, on the average, show a marked and even sudden dropping off in weight within two or three weeks after mating, which is not found in the case of the average single-bearing ewe.

This may be translated to mean *lowered nutrition after mating as a characteristic of the twin-bearing ewe;* and a consequent lessening of the growth of the unborn lambs, so that when the twin lambs are ultimately born they are of smaller size and weight than the average single lamb.

We sometimes find that twin lambs are very unequal in size at birth. If one is large the other is likely to be small and even deformed, indicating a struggle for existence between the unborn lambs.

This led me to suspect that many of our largest single lambs might be the survivors of twins; and a few years ago I had a young physician from Washington (Dr. Phelps) visit Beinn Bhreagh at the lambing season to see what he could discover bearing upon the point.

He was able to examine quite a number of the after-births of the sheep; and in several cases where single lambs had been produced he found indications in the after-birth of an aborted twin.

In this connection it is suggestive that our best twin-bearing ewe, No. 1505, which has given us already four sets of twins, produced this year a single lamb *weighing ten pounds at birth!* It is notorious also, that fat stock are remarkably infertile, and rarely bear twins.

All these considerations led to the belief that lessened nutrition during the period of gestation is a condition that is favorable to the preservation of unborn twins. Good nutrition at the time of mating favors the *conception* of twins; and diminished nutrition after mating favors their retention.

It is obvious, upon reflection, that ewes can not successfully bear twins, or a litter at birth, unless the lambs born are small; and lessened nutrition during the period of gestation is eminently favorable to the production of lambs that are small at birth.

These considerations led to the policy of giving the ewes extra nourishment in the form of oats, oil-cake, etc., for some time before mating; and then cutting off the extra feed soon after mating so as to throw the ewes back on the pasture alone for support.

A better, and certainly more economical plan, affecting the condition of nutrition in the same way, would be simply to mate in October when the pasture is at its best, and then give hay alone for winter feeding. The giving of oats, oil-cake, roots and other milk-promoting food might well be postponed until about the time of lambing, so as to avoid stimulating the growth of the lambs until after they are born.

In order to raise twins and triplets successfully the lambs should be small at birth, and grow large afterwards.

If we had a large number of twins from which to choose, it would be a good plan in the spring to retain only those lambs which are under the average weight at birth; and then, in the autumn, select from these those that are over the average weight. This process carried on through a series of generations would probably result in a breed of sheep producing, as a normal condition, lambs that are small at birth and which grow large afterward.

Alexander Graham Bell

Notes

1 From the *Beinn Bhreagh Recorder,* Vol. X., pp. 368–386: A typewritten periodical, limited to five copies, containing records of experiments of various kinds, conducted at Dr. Bell's summer place at Beinn Bhreagh, near Baddeck, Cape Breton Island, Nova Scotia.
One copy is deposited in the Smithsonian Institution, at Washington, D. C., and the others are in the possession of private individuals, viz., Dr. A. Graham Bell and Mr. J. G. Davidson at Beinn Bhreagh, N. S., and Mr. Gilbert H. Grosvenor and Mr. David C. Fairchild, at Washington, D. C.

2 See SCIENCE, Vol. XIX., p. 767.

3 This pamphlet may be found in many public libraries; the Library of Congress, Washington, D. C., Boston Public Library, the Library of the British Museum.

Part 7

PHYSIOLOGY

Editorial Headnote

Lydia Folger Fowler, *Familiar Lessons on Physiology: Designed for the Use of Children and Youth* (New York: Fowlers and Wells, 1848), I, 135–138

Lydia Folger Fowler, 1823–1879, was the second woman, after Elizabeth Blackwell, to gain a medical degree in the United States, going on to become a professor of medicine at Rochester Eclectic Medical College. Her text here is for children, and focuses on their powers of observation and perception, on the process of coming to know.

28

FAMILIAR LESSONS ON PHYSIOLOGY

Designed for the Use of Children and Youth

Lydia Folger Fowler

Source: New York: Fowlers and Wells, 1848, I, 135–138

OBSERVING, PERCEPTIVE, AND KNOWING FACULTIES

We have now come to the purely intellectual faculties. These are divided into two general classes, the Perceptive and Reflective Faculties. We shall first speak of the Perceptive Faculties. Their general use or object is to look, see, remember, collect facts and anecdotes, and remember them. I will first tell you about

24. INDIVIDUALITY.

Definition—*Observation; power of seeing and noticing objects; desire of looking at everything; curiosity.*
Location—*Individuality is situated at the root of the nose, in the lower part of the forehead, and, when large, gives fulness there.*

DOI: 10.4324/9781003009337-37

1. Individuality is one of the first organs that is developed in the mind of the child; and it is a wise order of nature that such is the fact; for the infant is ignorant of everything around him. It does not even know or recognize its own parents, and has no idea whatever of any object in creation. Its mind is like a blank book full of leaves, but without ideas and impressions.
2. What is the first thing that the little tiny tender thing does? It opens its little eyes, and looks, looks; it gazes at everything around it, and the little creature appears to be delighted when its attention is attracted to any bright object, as the light, or fire.
3. The child observes, and by observing it distinguishes its mother, and is conscious when she is present. The older it becomes, the more it looks and notices; and when Imitation has taught it to prattle and talk, it commences to ask questions, which is as natural to the child as to see.
4. You all know, children, that you can understand much better what flowers, birds, and animals are by seeing them, and when you receive instruction in school you know how much easier it is for you to understand, when your teacher shows you something which represents what she is talking about. Those who are born blind can learn a great deal, but they can comprehend but a very small portion of what they are deprived.
5. There was quite an aged gentleman who was so near-sighted, that he was unable to see the stars at night, some person procured a pair of spectacles for him, by which means his sight was so much improved that he could behold these worlds of beauty and light. His soul was filled with admiration, and he could never be satisfied in beholding them. He said that he never had an idea that they were half as beautiful.
6. Individuality is generally large in children, yet if it be not cultivated and strengthened by use, it will not be of much service. Two persons walk together in the street of a city; one will notice every house, every square, and every man and woman he meets, and will see everything that is to be seen; while the other only takes half a glance, and could not tell, the next time he passed through the same street, whether he had ever been there before or not. Travellers who visit the same countries give very different descriptions of the same places; because they have different degrees of Individuality, joined with other organs.
7. This organ sometimes gives curiosity to see and hear everything that transpires around us. It is this feeling of curiosity that probably disposes some children to be so meddlesome when they commence walking. They extend their hands, and put their little fingers on everything they see, and frequently do a great deal of mischief, while they are only seeing, seeing, looking, looking, just as they were designed to do.
8. The child crawls into the closet and pulls over the box of flour on its mother's clean carpet. This is amusement to the little one, and, unconscious of the mischief it has done, its little imagination has been exercised by thinking how beautifully the carpet was painted white.

9. You must not only look and observe, but you must learn to distinguish the difference between different things. Wherever you go, you must notice everything and use your Individuality.
10. If you should attend the museum, look distinctly at everything before you, and then ask questions about the birds, animals, and images. When you see an individual, look at his eyes, his hair, his nose, etc., and try to remember them. When you see a picture or engraving, observe every tree, figure, house, etc. In this way, and in this way only, you will gain much information, which will always be of service to you.
11. Of what use is your Benevolence if you do not discern the difference between true objects of charity and pretended beggars? You might give away all you possess, without benefiting those who needed alms. Of what use is your Constructiveness, if you do not notice the manner in which things are fitted and made? It will profit you nothing. Of what use is Ideality to you, unless you are delighted to observe the beauties and sublimities of nature? The rainbow would pass away before you noticed it. You could not appreciate the works of art, or of nature, merely because you do not see them.
12. In the cut, the gentleman is looking with his telescope at the stars. If he had no Individuality, he would not be able to perceive them, neither would he have any *inclination* to look at them.

Children, observe, look, take notice, and every day will add to your store of knowledge.

Notes

What is the subject of chapter seventh? Into what two classes are these divided? What is the use of the perceptive faculties? What is the definition of Individuality? What is its location?

1. Of what use is this organ to the child, and when is it developed?

2, 3. In what ways does it show this organ? 4. In what way can children best understand what they learn? 5. Relate the anecdote of the gentleman who was near-sighted?

6. In whom is this organ generally large? In what way do persons show the strength or deficiency of this organ? 7. What feeling does this organ give? How do children frequently gratify it? What is probably their motive? 8. Illustrate this by the child.

9. What must you do beside look and observe? 10. In what ways can much information be gained? 11. When would Benevolence be useless? Constructiveness? Ideality? 12. Explain the cut. What organ does he use? What general direction is given to children?

Editorial Headnote

Thomas Henry Huxley, *Lessons in Elementary Physiology* (London: MacMillan, 1868), vii–ix

Thomas Henry Huxley, 1825–1895, luminary of Victorian science, was principally a biologist, but had an enormous influence in scientific teaching and government policy regarding science across the board. Known as 'Darwin's Bulldog' for his vociferous advocacy of evolution, he is also known as the coiner of 'agnosticism' as expressing a limit of what can be both proven and denied according to its capacity to be tested. His *Lessons in Elementary Physiology* landed Huxley in trouble with scientific critics, who thought he was corrupting the youth with instructions on unnecessary repetition of painful vivisection (which Huxley himself did not have the stomach for). The emphasis is on the need for the production of expertise through direct experience, meaning that students had to experiment for themselves. This edition contains the original preface and its update.

29

LESSONS IN ELEMENTARY PHYSIOLOGY

Thomas Henry Huxley

Source: London: MacMillan, 1868, vii–ix

PREFACE TO THE SECOND EDITION

THE present edition of the "Lessons in Elementary Physiology," has been very carefully revised. A few woodcuts have been added; others have been replaced by better ones, as in the case of the figures of the retina, which embody the results of Schultze's latest researches.

Some additions (but as few as possible, lest the book should insensibly lose its elementary character) have been made; among the most important I count the very useful "Table of Anatomical and Physiological Constants" drawn up for me by Dr. Michael Foster, for whose friendly aid I am again glad to express my thanks.

It will be well for those who attempt to study Elementary Physiology, to bear in mind the important truth that the knowledge of science which is attainable by mere reading, though infinitely better than ignorance, is knowledge of a very different kind from that which arises from direct contact with fact; and that the worth of the pursuit of science as an intellectual discipline is almost lost by those who seek it only in books.

As the majority of the readers of these Lessons will assuredly have no opportunity of studying anatomy or physiology upon the human subject, these remarks may seem discouraging. But they are not so in reality. For the purpose of acquiring a practical, though elementary, acquaintance with physiological anatomy and histology, the organs and tissues of the commonest domestic animals afford ample materials. The principal points in the structure and mechanism of the heart, the lungs, the kidneys, or the eye, of man, may be perfectly illustrated by the corresponding parts of a sheep; while the phenomena of the circulation, and many of the most important properties of living tissues, are better shown by the common frog than by any of the higher animals.

Under these circumstances there really is no reason why the teaching of elementary physiology should not be made perfectly sound and thorough. But it should be remembered that, unless the learner has previously acquired a knowledge of

DOI: 10.4324/9781003009337-38

the elements of Physics and of Chemistry, his path will be beset with difficulties and delays.

PREFACE TO THE FIRST EDITION

THE following "Lessons in Elementary Physiology" are primarily intended to serve the purpose of a text-book for teachers and learners in boys' and girls' schools.

My object in writing them has been to set down, in plain and concise language, that which any person who desires to become acquainted with the principles of Human Physiology may learn, with a fair prospect of having but little to unlearn as our knowledge widens.

It is only by inadvertence, or from an error in judgment, therefore, that the book contains any statement, or doctrine, which cannot be regarded as the common property of all physiologists. I have endeavoured simply to play the part of a sieve, and to separate the well-established and essential from the doubtful and the unimportant portions of the vast mass of knowledge and opinion we call Human Physiology.

Editorial Headnote

Thomas Henry Huxley, *Lessons in Elementary Physiology* (London: MacMillan, 1885), v–vi

In this later edition of the same work, Huxley makes specific if oblique mention of the fact that earlier versions of the work were misconstrued as encouraging cruelty. At the same time, Huxley acknowledges the open-endedness of the empirical method. This work, two decades after it first appeared, is vastly revised.

30

LESSONS IN ELEMENTARY PHYSIOLOGY

Thomas Henry Huxley

Source: London: MacMillan, 1885, v–vi

PREFACE TO REVISED EDITION, (1885)

IT is now more than twenty years since I planned and began to write these "Lessons," the object of which is defined in the preface to the first edition.

During this period physiological investigations have displayed an activity unprecedented in history. Here, as in all such branches of natural knowledge, the method of experiment has shown itself to be the one path which leads to truth, and has not only revealed multitudes of novel physiological facts, but has suggested wholly new lines of inquiry.

As successive editions of the "Lessons" have been demanded, it has been my effort to incorporate with them such solid acquisitions of the ever advancing science of physiology as fall within their scope, while rigorously excluding all debatable matter or, at any rate, carefully indicating it as such. At the same time, long experience as a Teacher and Examiner having forcibly brought home to my mind the difficulty of making any statement that cannot be misunderstood, an amount of attention has been devoted to questions of mere exposition, which really deserves, but probably has not attained, the reward of abolishing such misunderstandings.

The present edition has been more extensively revised than any of its predecessors. The chapter on Histology, in particular, has been entirely reconstructed and provided with new illustrations, several of which have been taken from Ranvier and from Quain.

In the preface to former editions of this work, I have had the pleasure of expressing my obligations to Dr. Foster, Secretary to the Royal Society, for his very valuable assistance. That aid has been still more freely rendered to the present edition, which, in fact, could not have appeared unless Dr. Foster had taken upon himself the whole burden of seeing the work through the press. My friend has indeed done so much during my enforced idleness, that I should have been better satisfied if he would have permitted me to associate his name with mine on the title-page.

DOI: 10.4324/9781003009337-39

Editorial Headnote

Henry Head and William Halse Rivers, 'A Human Experiment on Nerve Division', *Brain* (1908), 324–342

Henry Head, 1861–1940, was a neurologist.

William Halse Rivers, 1864–1922, was a neurologist, anthropologist and psychiatrist.

Head was particularly noted for using his own body as a site of experimentation, as here, showing an exceptional practical consequence of the core scientific self's need to know and know with certainty.

31

'A HUMAN EXPERIMENT ON NERVE DIVISION'

Henry Head and William Halse Rivers

Source: *Brain* (1908), 324–342

CHAPTER I—HISTORY OF THE CASE

It had long been recognized that the consequences of injury to a peripheral nerve could not be adequately explained on any accepted theory of its structure and function. In 1901, Dr. Head and Mr. Sherren therefore determined to make a systematic examination of the patients attending the London Hospital for some nerve injury. The hospital patient is frequently an admirable subject for sensory experiments; at his best he answers "Yes" and "No" with certainty, and is commendably steady under the fatigue of control experiments. Moreover, the number of patients, who come to the London Hospital for such injuries, is so large that it is possible to eliminate entirely those who are found to be untrustworthy in consequence of misuse of alcohol or other causes.

Most of the main facts of nerve distribution and recovery of sensation can be elicited from a study of hospital patients by means of simple tests requiring no undue expenditure of time. But such patients can tell little or nothing of the nature of their sensations, and the time they are able, or willing, to give is insufficient for elaborate psycho-physical testing.

It soon became obvious that many observed facts would remain inexplicable without experimentation carried out more carefully and for a longer period than was possible with a patient, however willing, whose ultimate object in submitting himself to observation is the cure of his disease. For instance, an examination of the part played by heat- and cold-spots in the return of sensation was impossible under clinical conditions.

It is also unwise to demand any but the simplest introspection from patients, to whatever class they may belong. This side of the investigation was, therefore, almost entirely closed to Mr. Sherren and Dr. Head. From the early days of their research, Dr. Rivers had acted as their guide and counsellor. His interest lay rather in the psycho-physical aspect of the work, and he was impressed with the insecurity of this side of the investigation. Introspection could be made fruitful by the personal experiences of a trained observer only.

DOI: 10.4324/9781003009337-40

Lastly, we were anxious to investigate the functions of deep sensibility. Sherrington [25] had shown that muscular nerves contained a large number of afferent fibres. From the beginning of their research, Head and Sherren [17] had tried to determine the sensibility remaining after complete division of all cutaneous nerves without injury to the muscular branches. But accidental injuries of this kind are excessively rare, and they were compelled to attack the problem by indirect and less satisfactory methods. As soon, therefore, as it was determined to make an experimental division of peripheral nerves, means were taken to ensure that the nature of these deep afferent fibres should come clearly to experimental investigation.

At the time of the experiment, H. was nearly 42 years of age and in perfect health. Since boyhood he had suffered from no illnesses, excepting as the consequence of wounds in the *post-mortem* room. None of these had attacked his left arm or hand, which were entirely free from scars or other deformities.

For two years before these experiments began he had given up smoking entirely. No alcohol was ever taken on the days during which he was under examination, and for some years he had abstained from alcohol except on holidays.

On April 25, 1903, the following operation was performed by Mr. Dean,[1] assisted by Mr. Sherren.

An incision 6½ in. (16·5 cm.) long was made in the outer bicipital fossa extending along the axial line of the left upper extremity; this wound was almost exactly bisected by the fold of the elbow. After turning back the skin, the supinator longus was hooked outwards, and the radial nerve (ramus superficialis nervi radialis) was divided at the point where it arises from the musculo-spiral (N. radialis). A small portion was excised, and the ends united with two fine silk sutures. The external cutaneous nerve (N. cutaneus antibrachii lateralis) was also divided where it perforates the fascia, above the point where its two branches are given off to supply the anterior and posterior aspects of the pre-axial half of the forearm. The nerve was sutured with fine silk, and the wound was closed with silk sutures, without drainage. The limb was put up on a splint with the forearm flexed at the elbow, and the whole hand was left free for testing. The wound healed by first intention.

The following morning (April 26, 1903), the radial half of the back of the hand and dorsal surface of the thumb were found to be insensitive to stimulation with cotton wool, to pricking with a pin, and to all degrees of heat and cold. Around the base of the index and middle fingers was a small area insensitive to stimulation with cotton wool and von Frey's hairs, where a response was obtained to the prick of a pin. No sensation was evoked by any manipulation of the hairs within the affected parts on the back of the hand.

The area insensitive to cotton wool extended slightly further towards the ulnar aspect of the back of the hand than that of the cutaneous analgesia. Between the two lay a narrow zone, where a painful cutaneous stimulus produced a more unpleasant sensation than over the normal skin.

The most striking fact, however, was the maintenance of deep sensibility over the whole of the affected parts on the back of the hand. Pressure with the finger, with a pencil, or any blunt object was immediately appreciated. All those stimuli commonly used by the clinician to test the presence of "touch" were appreciated and well localized. Mr. Dean, who was not familiar with our previous observations, said he should have thought that sensation of touch was intact, had he not known the nerves had been divided.

On May 4, nine days after the operation, the hand was exposed to a long series of experiments. The most striking features of this examination were:—

(*a*) That very moderate pressure on the abnormal area of the skin was appreciated and could be well localized, whilst touches with cotton wool, or deformations of the skin, produced by drawing the hair outwards, caused absolutely no sensation.
(*b*) In spite of the existence of this sensibility, two compass-points could not be distinguished, even when separated by 8 cm.
(*c*) All sensation was lost to cutaneous painful stimuli, and to heat and cold. In fact, the condition might easily have been mistaken for one of analgesia and thermo-anæsthesia with intact sensibility to touch.
(*d*) Between the extent of the analgesic area and that insensitive to cotton wool, lay a border where the prick of a pin was abnormally painful.
(*e*) None of the cold-spots marked out before the operation reacted to the usual stimuli.

By May 4 the skin on the back of the hand had assumed a peculiar condition, which was described on the 7th by Dr. J. H. Sequeira in the following words:—

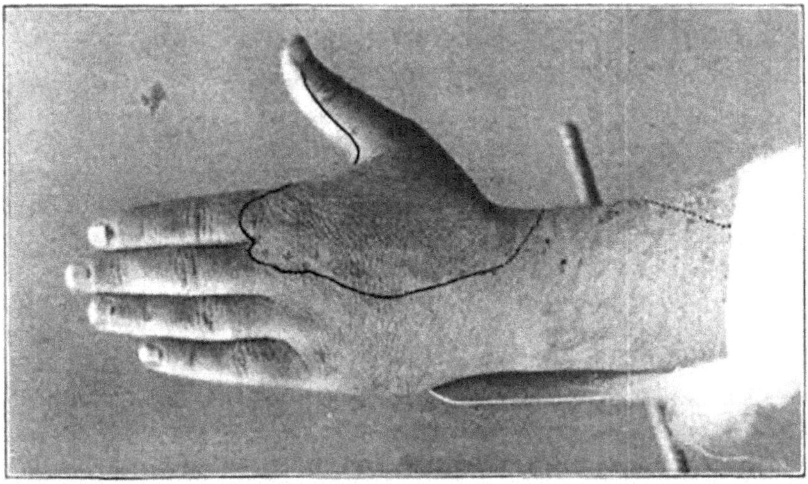

Figure 31.1 To show the extent of the loss of sensation produced by the operation.
The anæsthesia to cotton wool and to von Frey's hairs is bounded by the black line. The analgesia to prick and other cutaneous painful stimuli lay within the red crosses.
The darkness of the affected area is due to its deep red colour compared with the rest of the hand.

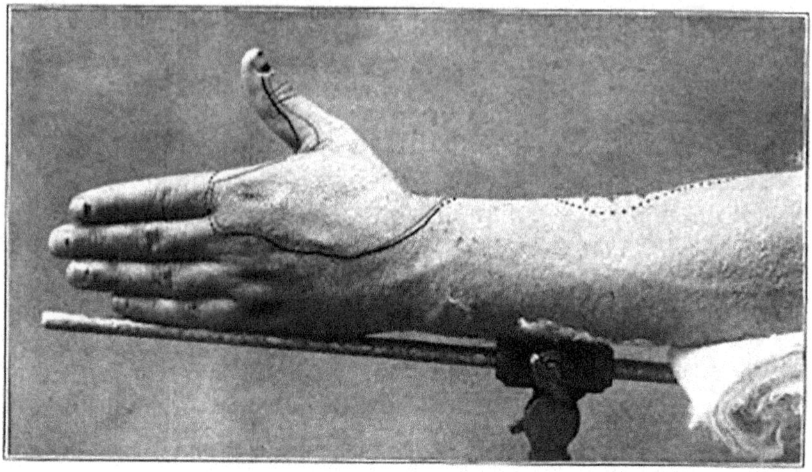

Figure 31.2 To show the extent of the loss of sensation twenty-one days after the operation (May 16, 1903). The black line encloses the loss to cutaneous tactile stimuli; the red line encloses the cutaneous analgesia. Wherever these lines are broken the border was an indefinite one.

"The whole of the affected area is of a slightly deeper red than the rest of the skin of the hand. It is dry, and covered with minute hair-like scales. On palpation, the skin appears to be thickened and looks as if it were slightly œdematous; but it does not pit on pressure. A striking feature is the absence of the normal elasticity, which is in remarkable contrast with the rest of the skin. The affected parts do not sweat, while the rest of the hand is permanently slightly moist."

From the time of the operation until the removal of the splint (May 23), the borders of the loss of sensation on the forearm underwent no material change. But H.'s skin has always been peculiarly susceptible to the action of chemical antiseptics, and the necessary cleansing at the time of the operation led to desquamation to within about 3 in. to 4 in. of the wrist. Fortunately, the hand had entirely escaped their action.

On the extensor aspect of the forearm, the loss of all forms of cutaneous sensation was bounded for the greater part by a definite line. Towards the radial aspect, the loss of sensation merged more gradually into parts of normal sensibility. The borders formed a sinuous line, seen on figures 31.3 and 31.4. Over the greater part of the forearm, the loss of sensation to prick was less extensive than that to cotton wool; but nearer to the wrist, the reverse condition seemed to exist.

The extent of the cutaneous analgesia on the hand was slightly less than that of the loss of sensation to cotton wool and von Frey's hairs, and to these stimuli all the boundaries were sharply defined, except at the base of the index and middle fingers.

The splint was removed on May 23, and it was then possible to wash the arm vigorously and to remove the loose scales of epithelium. We then discovered that

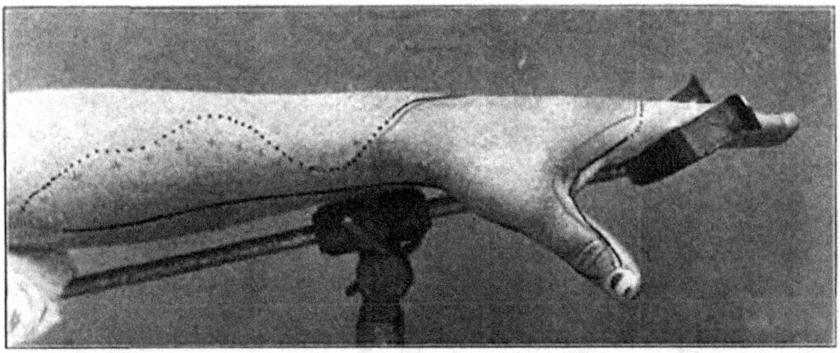

Figure 31.3 Lateral view on the same date (May 16, 1903).

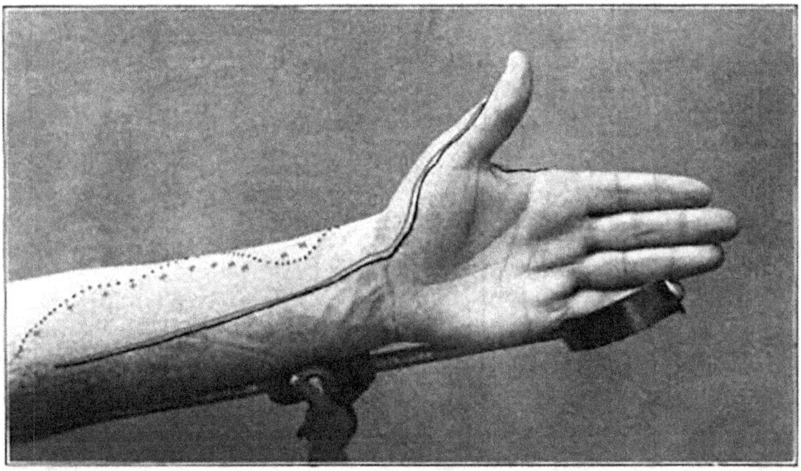

Figure 31.4 Flexor aspect of the forearm on the same date (May 16, 1903).

the loss of sensation to prick was everywhere coterminous with, or slightly less extensive than, the loss to cotton wool, except near the wrist. Here there was a triangular area, shown on figure 31.5, where cotton wool and No. 5 of von Frey's hairs were undoubtedly appreciated, although the skin was insensitive to prick.

On the back of the hand, sensibility remained exactly in the condition described immediately after the operation. Over the whole area of cutaneous anæsthesia, pressure-touches were appreciated and well localized. Pain could be produced as easily by pressure with the algometer over the back of the affected as over similar parts of the normal hand. Electrical stimuli produced no sensation except when the muscles contracted; then the smallest visible movement was appreciated. To recognize pure movement, produced electrically, without a concomitant cutaneous sensation is a remarkable experience.

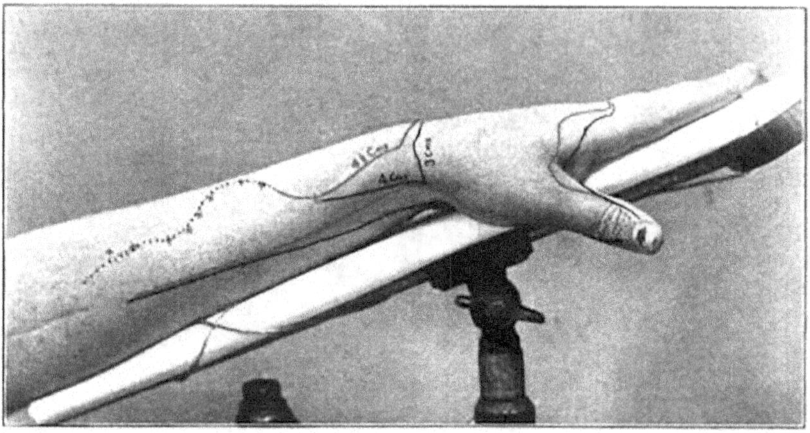

Figure 31.5 To show the loss of sensation on May 26, 1903 (thirty-one days after the operation). On the lateral aspect of the forearm near the wrist is shown the triangular area insensitive to prick and other cutaneous painful stimuli but sensitive to stimulation with cotton wool (*vide* p. 400).

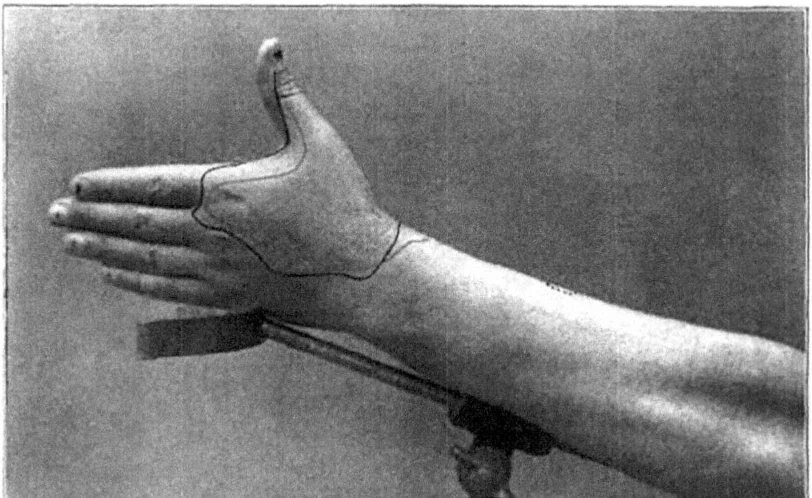

Figure 31.6 To show the loss of sensation on June 14, 1903 (fifty days after the operation).

Though sensitive to the tactile and painful elements of pressure, and to the passive movement of muscles, the back of the hand was anæsthetic to all thermal stimuli; the tissues could be frozen firmly with ethyl chloride without the production of even the slightest sensation.

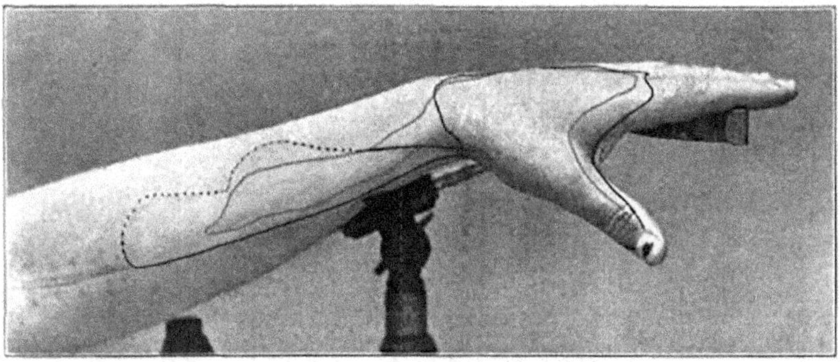

Figure 31.7 Lateral view of the forearm and hand on June 14, 1903.

The first noticeable change in the extent of the loss of sensation was discovered on June 7, forty-three days after the operation. The borders of the area insensitive to cotton wool remained unaltered, but the cutaneous analgesia was distinctly less extensive, and no longer coincided with it on the flexor aspect of the forearm; the extent of the cutaneous analgesia had diminished for 3 in. (8 cm.), or more, peripheral to the scar. This was particularly noticeable, because the borders of the loss of sensation to cotton wool had remained unchanged.

Moreover, the boundaries of the cutaneous analgesia were no longer definite, but were made up of islets, or points, of sensation. Passing from the area of complete loss to parts normally sensitive to a prick, the pin struck spots, where it produced a slowly developed, dull but painful sensation. If, however, this particular spot was not struck, it might be that sensibility was not encountered until another spot was reached, some millimetres nearer the unaffected parts of the limb.

On June 14, fifty days after the operation, the gradual shrinking in extent of the cutaneous analgesia on the arm was found to have continued (figures 31.6, 31.7 and 31.8), although the borders of the loss of sensation to cotton wool remained entirely unaltered. The loss of sensation to cold corresponded in extent with that of the loss to prick; but, wherever the part was feebly sensitive to the latter stimulus, sensibility to cold seemed to be absent. To all degrees of heat the borders of the loss of sensation had remained unchanged, and the extent of the anæsthesia, even to temperatures between 50° C. and 60° C., uniformly exceeded that of the loss to prick.

By June 20 (fifty-six days after the operation), recovery of sensation had progressed still further. Not only had the extent of the absolute cutaneous analgesia shrunk considerably on the forearm, but the terminal phalanx and a portion of the basal phalanx of the thumb had become sensitive to prick. The extent of the area between the borders of the loss of sensation to cotton wool and to prick in the first interosseous space had increased to nearly 2 cm. (figure 31.9). At this date, the markings of the previous week were still visible on the arm, and the boundary of the analgesia lay in many places 0·5 cm. within those determined six days before. And yet, in spite of this rapid

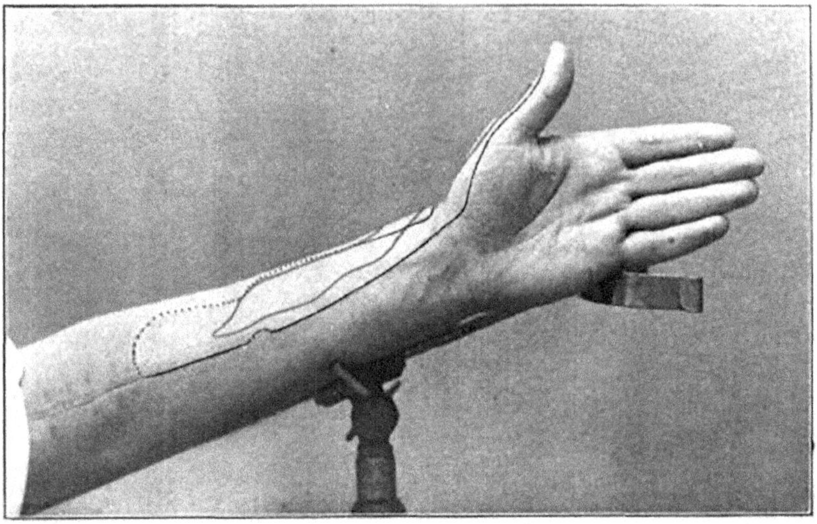

Figure 31.8 Flexor aspect of the forearm on June 14, 1903.

improvement in sensibility to prick, the borders of the anæsthesia to cotton wool and von Frey's tactile hairs remained absolutely unchanged.

On the flexor surface of the forearm, there was nothing to show that cold could be appreciated within the border of cutaneous tactile anæsthesia. In the first interosseous space, cold was certainly appreciated well inside the limits of the loss of sensation to cotton wool; the border of the loss to cold lay about midway between that for cotton wool and that for prick.

By July 20 (eighty-six days after the operation), there was no part of the forearm where a prick could not be occasionally appreciated, although in many places this form of sensation was extremely defective. Moreover, considerable changes had occurred in the condition of the hand; the whole of the thumb and the skin over the radial half of the first metacarpal had become sensitive to prick. The analgesic area on the back of the hand was diminishing from its radial aspect.

In spite of these changes, the borders for the loss of sensation to cotton wool remained exactly as before.

On the forearm, ice was not appreciated with certainty, until the original border of cutaneous analgesia was passed. Water above 50° C. produced pain within the parts now sensitive to a prick, but it was impossible to say whether the pain was accompanied by any thermal quality.

The terminal phalanx of the thumb was certainly sensitive to cold below 17° C. and more doubtfully to heat above 45° C. Within the area of dissociated sensibility in the first interosseous space, and over the ball of the thumb, it was difficult to be sure that any sensation of temperature was produced by ice-cold and hot water; but the border of the thermo-anæsthesia probably lay slightly within that for sensibility to cotton wool.

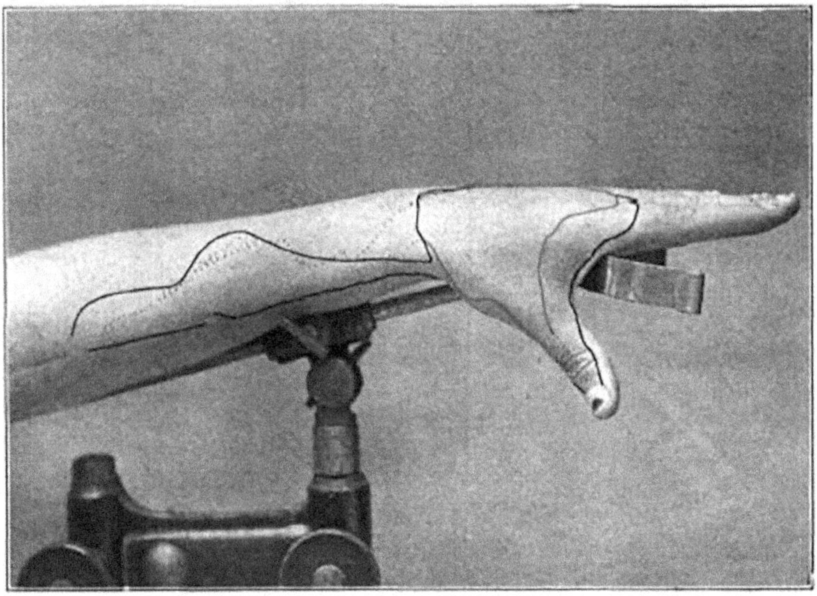

Figure 31.9 To show the loss of sensation on June 20 (fifty-six days after the operation). The cutaneous analgesia on the forearm could not now be defined accurately, but merged everywhere into parts sensitive to prick. It is therefore surrounded by a dotted broken border.

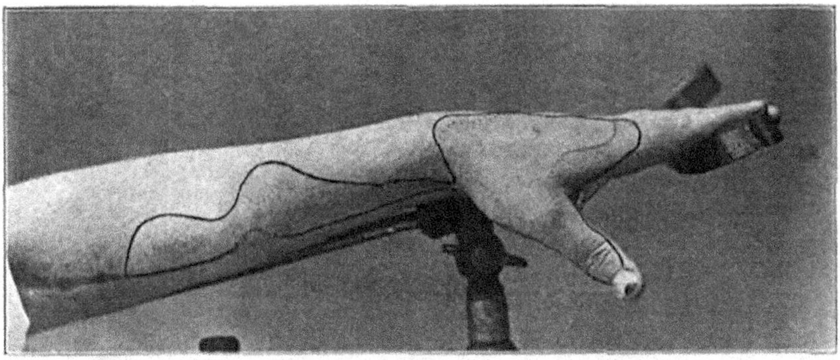

Figure 31.10 To show the condition on July 20, 1903 (eighty-six days after the operation). No part of the forearm was entirely insensitive to cutaneous painful stimuli.

At this time, some of the hairs on the forearm within the affected area became sensitive to pulling. The sensation produced was slowly developed and excessively unpleasant. It died away, and recurred again, without further stimulation. Those hairs lay entirely within the upper anæsthetic patch.

On August 10 (107 days after the operation), the sensibility to prick had further improved, although the extent of the anæsthesia to cotton wool remained entirely unaltered. On August 15, for the first time since the recovery of sensation began, it could be said that parts which were at first insensitive to heat and cold now responded definitely and constantly to these stimuli. Over the upper patch on the forearm, ice uniformly produced a sensation of cold. Temperatures above 50° C. caused a stinging sensation, usually called "burning," but it is doubtful to what extent this contained more than the painful element of heat.

The terminal phalanx of the thumb undoubtedly responded to temperatures above 45° C., and the sensation produced by temperatures above 50° C. contained a thermal element in addition to the stinging pain. Even the proximal phalanx of the thumb had become sensitive to ice, although still anæsthetic to heat.

On August 15 and 16, these observations on the upper patch of the forearm and the terminal phalanx were confirmed, and within these areas we were able to mark out definite cold-spots for the first time since the operation. Four of these lay in the upper patch, and four over the terminal phalanx of the thumb.

By September 9 (137 days after the operation), the whole forearm had become sensitive to cold, and cold-spots were discovered not only in the upper patch on the forearm and in the terminal phalanx of the thumb, but also over the more distal portions of the affected area. The forearm still remained insensitive to heat, except in as far as temperatures above 50° C. produced a peculiar form of painful sensation, usually called "burning."

In spite of the complete absence of any change in the behaviour of the affected parts to cotton wool, sensibility to prick continued to return steadily, and by September 24 (152 days after the operation) a small area only on the back of the hand remained insensitive to this stimulus (figure 31.11).

Since July 10, when the back of the hand had been too energetically frozen with ethyl chloride, an indolent sore had existed in the centre of the affected area.

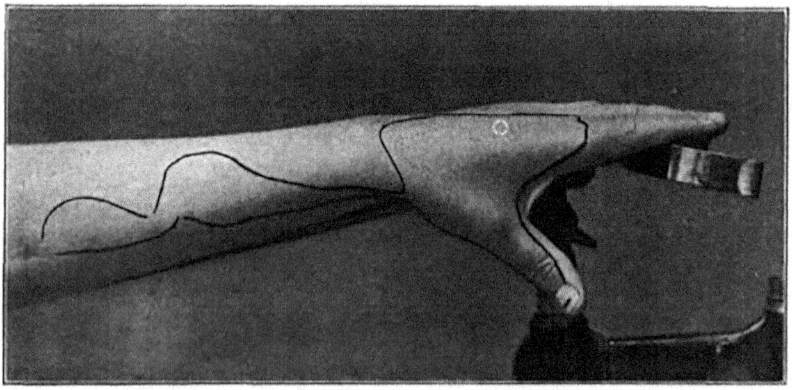

Figure 31.11 To show the condition on September 24, 1903 (152 days after the operation). Over the small irregular area on the back of the hand sensibility to cutaneous painful stimuli was greatly diminished and in places was absent.
The area occupied by the trophic sore is marked with a circle.

It tended to heal if protected, but would break down again under the influence of the slight accidents of ordinary life. On September 23, attention was attracted to its condition by the presence of tingling, which had never been noticed before. This led to the discovery that painful sensation, of a dull and ill-defined character, was present in the neighbourhood of the sore. From this time it healed with great rapidity, although no special care was taken to protect it. Once healed, it never broke down again after the return of sensibility to painful cutaneous stimuli.

About this time, part of the first interosseous space, which had become sensitive to prick, began to respond to ice; this return of sensation was found to be associated with a few definite cold-spots. But the affected area still remained insensitive to heat.

On October 3 (161 days after the operation), we noticed, for the first time, that cotton wool produced some sensation over the upper patch on the forearm. This change advanced with considerable rapidity, and on October 6 sensibility to cotton wool was present in a very defective form over both upper and lower forearm patches. The upper of these areas seemed to become sensitive by gradual encroachment from the edges, whilst the lower appeared to recover at the centre as quickly as at the periphery. Later we found that this response was due entirely to the return to the hairs of a peculiar form of sensibility (*vide* p. 385).

About this time (October 8), the upper patch on the forearm became undoubtedly sensitive to temperatures of and above 45° C. An excellent heat-spot was found in the centre of the patch, to which this return was certainly due.

By October 15 (173 days after the operation), no part of the hand was entirely insensitive to prick, although sensation was defective over the parts dotted on figure 31.12.

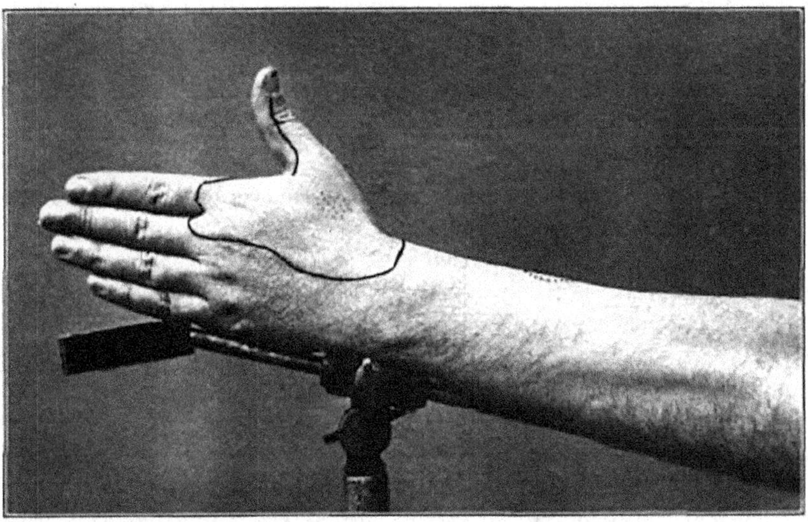

Figure 31.12 To show the condition of the back of the hand on October 15, 1903 (173 days after the operation). The healed sore can be seen as a scar in the neighbourhood of the red dots.

The greater part of the back of the hand now reacted to the more extreme degrees of cold, and the cold-spots had multiplied greatly. By November 1 (190 days after the operation), cold could be appreciated everywhere over the back of the hand, and twenty-four cold-spots were discovered within the affected area. At the same time, one heat-spot was found near the base of the first phalanx of the thumb. This was the only part of the affected area on the hand sensitive to heat.

From this time, the cold-spots and heat-spots rapidly increased in number over the back of the hand, the increase proceeding step by step with the recovery of sensibility to cold and to heat.

With the gradual return of sensibility to pain, cold, and heat, we noticed that the sensation tended to be widely diffused, and was not infrequently localized in some part remote from the point of stimulation (September, 1903). If, for instance, ice was applied to the proximal portion of the forearm, a sensation of coldness was produced in the thumb. The site of this referred sensation remained the same, whatever the nature of the stimulus, provided it was one to which the affected area had become sensitive.

By December 3 (222 days after the operation), the peculiar tingling sensation produced by cotton wool could be evoked by stimulating the thumb and the adjoining interosseous space. This sensibility rapidly increased in extent, until there was scarcely any part of the affected area from which it could not be produced (December 6). The sensation was one which could be expressed only as a general state of diffuse painless tingling. Moreover, it was found that parts which gave this reaction to cotton wool were insensitive to No. 5 of von Frey's hairs and to the painless interrupted current, just as in the early days after the operation. Exactly the same borders could be marked out both on the forearm and hand by dragging a pin lightly from normal to abnormal parts; for as soon as the old border of cutaneous anæsthesia to touch was passed, the sensation became a widely diffused tingling pain.

Thus it would seem that the sensibility to cotton wool, which began to return to the forearm 161 days, and to the hand 224 days, after the operation, was not the equivalent of the normal sensation of light touch over hairless parts, but was a peculiar form of hair-sensibility. For the areas endowed with it remained anæsthetic to the painless interrupted current and to No. 5 of von Frey's hairs; moreover, the sensation produced was widely diffused and was referred to remote parts, exactly like the sensation of prick and ice-cold over the same regions. This hypothesis was found at a later date to be correct. For on shaving the areas endowed with this form of sensibility, they became entirely insensitive to cotton wool.

We could not be certain that the forearm was sensitive to cotton wool when carefully shaved, until April 24, 1904, exactly a year after the operation.

On June 5, 1904 (407 days after the operation), the affected area on the forearm responded to temperatures of 37° C. This sensibility to warmth rapidly increased, and on June 26 was obtained, even with 34° C. Moreover, the sensation produced was one of warmth localized in the part touched. Except that it was not quite so acute, it exactly resembled that produced on the normal skin under similar circumstances. It had none of the diffuse radiation and tendency to reference into remote parts, so characteristic of the sensation evoked by stimulating heat-spots.

It was not until November 12, 1904 (567 days after the operation), that a portion of the back of the hand (figure 31.13) began to be undoubtedly sensitive to warmth (35·5° C.) and to cotton wool after shaving. The diffusion and radiation so characteristic of the previous stage of recovery were at once greatly diminished; so profound was this change that we recognized it before we could be certain of the increased sensibility to thermal and tactile stimuli.

Up till the end of November, 1904, the improvement continued rapidly. But with the advent of winter cold the condition of the hand went back; the referred sensations reappeared, to become as definite as they had been six months before, and the hand became less sensitive to warmth and to cotton wool after shaving.

In March, 1905, it again began to improve. Part of the affected area on the back of the hand became sensitive to No. 5 of von Frey's hairs, and reference greatly diminished. By May 21, a large area on the back of the hand had become sensitive to cotton wool and to minor degrees of heat; stimulation with No. 5 was widely appreciated. A referred sensation could no longer be produced from any part in this condition.

Although this improvement continued throughout the summer of 1905, a small portion of the affected area, lying mainly in the neighbourhood of the knuckles of the index and middle finger, has remained insensitive to No. 5, or to cotton wool after shaving (figure 31.14). Even at the present time, this part still is in a purely protopathic condition, sensitive to prick, to ice and to water above 37° C.

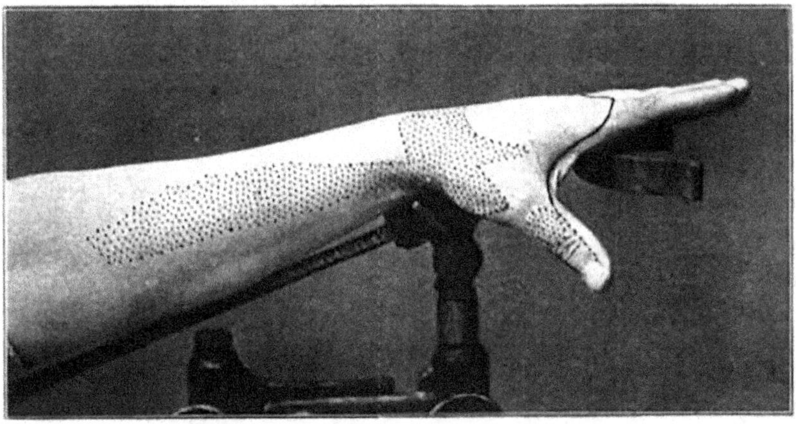

Figure 31.13 November 12, 1904 (567 days after the operation).

To show the manner in which sensibility returned to cutaneous tactile stimuli. The dotted area corresponds to the parts sensitive after shaving to cotton wool and to von Frey's tactile hairs (No. 5). These parts were also sensitive to temperatures of about 36° C.

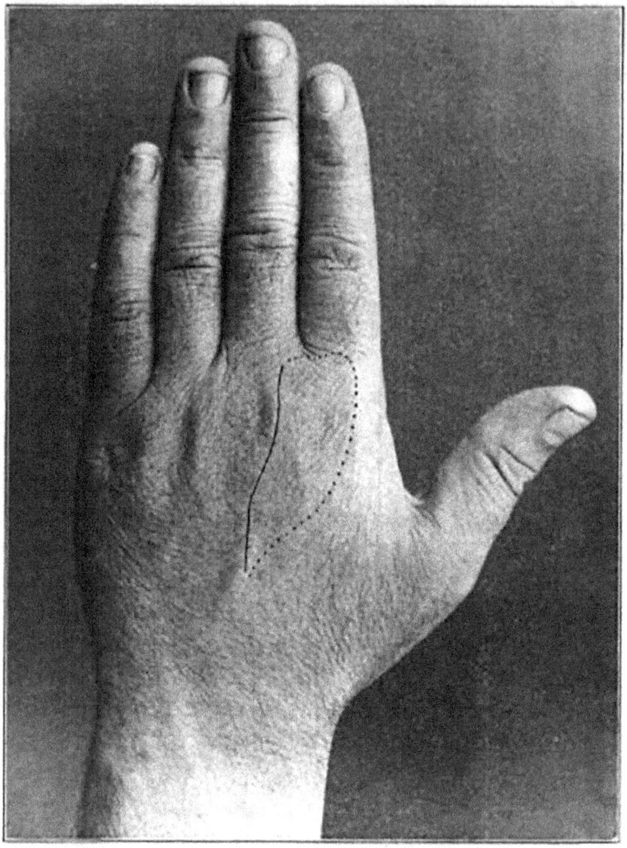

Figure 31.14 To show the extent of the affected area, which is still (1908) supplied with deep and protopathic sensibility only. Its radial border merges gradually into parts that have recovered more completely.

All these stimuli cause sensations, referred to the dorsal aspect of the thumb and diffused widely around the point to which they are applied. It seems as if one of the branches of the external cutaneous had not reunited, leaving this part of the affected area to be supplied by its fellow and by the radial.

The history of the case may be summed up in the form of the following diary: –

On April 23, 1903, the radial (ramus superficialis nervi radialis) and both branches of the external cutaneus (N. cutaneus antibrachii lateralis) were divided in the neighbourhood of the elbow. Both nerves were reunited with silk sutures and the wound healed by first intention.

This operation did not interfere with sensibility to the tactile and painful aspects of pressure. But the whole of the affected area became insensitive to prick, to heat, and to cold; two points of the compasses, applied simultaneously, could not be appreciated, but localization was preserved.

Forty-three days after the operation (June 7), the extent of the cutaneous analgesia had begun to diminish.

Fifty-six days after the operation (June 20), the analgesia on the forearm had greatly diminished, and the thumb had become sensitive to prick.

Eighty-six days after the operation (July 20), the whole forearm responded to prick, and the back of the hand was becoming rapidly sensitive to this form of stimulation. Cold was not appreciated except over the terminal phalanx of the thumb, and 50° C. gave rise to no sensation of heat.

One hundred and twelve days after the operation (August 15), the proximal part of the affected area over the forearm had become sensitive to cold.

One hundred and thirty-seven days after the operation (September 9), the whole forearm had become sensitive to cold.

One hundred and fifty-two days after the operation (September 24), the whole of the affected area, excepting a small spot on the back of the hand, had become sensitive to prick; the trophic sore healed.

One hundred and sixty-one days after the operation (October 3), cotton wool began to produce a diffuse tingling sensation over the forearm when the hairs were stimulated, but the whole of the affected area still remained insensitive to von Frey's tactile hairs. About the same time, the proximal patch on the forearm began to be sensitive to heat, and a definite heat-spot was discovered in this position.

One hundred and seventy-three days after the operation (October 15), the whole of the back of the hand had become sensitive to prick and, in a less degree, to cold.

One hundred and ninety days after the operation (November 1), the first heat-spot was discovered on the back of the hand.

Two hundred and twenty-five days after the operation (December 6), the hairs on the back of the hand responded with a diffused tingling to cotton wool, but the whole affected area of the forearm and hand still remained insensitive to von Frey's tactile hairs. This sensibility to cotton wool disappeared at once, if the arm was carefully shaved.

Three hundred and sixty-five days after the operation (April 24, 1904), the proximal patch on the forearm began to be sensitive to cotton wool after shaving.

Between *four hundred and seven* and *four hundred and twenty-eight days* after the operation (June 5 to June 26, 1904), the affected area on the forearm became sensitive to temperatures between 37° C. and 34° C. The tendency to diffusion and reference greatly diminished.

Five hundred and sixty-seven days after the operation (November 12, 1904), the greater part of the affected area on the back of the hand had become sensitive to cutaneous tactile stimuli, and temperatures below 37° C. evoked sensations of warmth.

Note

1 To Mr. Dean our best thanks are due, not only for the exactitude with which he carried out our wishes, but also for his kindness in receiving Dr. Head into his house for the operation.

Editorial Headnote

Edward Sharpey Schafer, *Experimental Physiology* (London: Longmans, Green, and Co., 1912), vi, 1

Edward Albert Sharpey-Schafer, 1850–1935, was a physiologist. He added the name Sharpey to honour his doctoral supervisor, William Sharpey. One of the most prominent of the original generation of professional physiologists, he is famed for the discovery of adrenaline and for having founded endocrinology. The excerpt here is short, emphasising both the basis of instruction in his own experience, and listing explicitly the kind of experimental experience required by any student before beginning a course in physiology.

32

EXPERIMENTAL PHYSIOLOGY

Edward Sharpey Schafer

Source: London: Longmans, Green, and Co., 1912, vi, 1

PREFACE

THE present volume is based upon a small book, entitled *Directions for Class Work in Practical Physiology*, which was published in 1901, and consisted chiefly of an account of the methods employed in the study of muscle and nerve physiology and of the heart, in so far as these could be illustrated in preparations from the frog. To that account have now been added descriptions of such other methods of Experimental Physiology as may be performed by the student either in class or individually, although a few which are given can—at least in this country—only take the form of demonstrations. The additions deal chiefly with the heart and vascular system, respiration, secretion, the central nervous system, and the sense organs; but considerable changes have also been introduced in the description of the methods of studying the physiology of muscle and nerve, and not only have a large number of new illustrations—for the most part in the form of diagrams—been added, but all the original figures have been redrawn, so that the work has taken the character of a new book, which it has been thought proper to furnish with another name.

The descriptions of the methods employed are founded upon twelve years' experience in conducting the classes of Practical Physiology in the University of Edinburgh. These classes are of two kinds, elementary and advanced, and some of the methods described are intended only for advanced students. But no attempt has been made in the work to discriminate between the experiments which are expected to be performed by the one or the other grade. Such selection can be best made by the teacher of the subject, and would be regulated by the requirements of the students and the nature of the examinations for which they are preparing.

EXPERIMENTAL PHYSIOLOGY

Preliminary

As a preliminary to any course of Experimental Physiology, the following dissections should be made:—

DOI: 10.4324/9781003009337-41

Frog. In this the following parts are to be displayed:—

(*a*) The brain and spinal cord, with the nerves passing off from them.
(*b*) The heart, with the main vessels leading to and from it.
(*c*) The lungs.
(*d*) The abdominal viscera, including stomach, small intestine, large intestine, liver, spleen, kidneys, bladder, and reproductive glands.
(*e*) The nerves and muscles of the leg.

Rabbit.

(*a*) The skull to be opened, and the several parts of the brain examined.
(*b*) The trachea and the blood-vessels and nerves on either side of it to be displayed, and to be followed as far as possible both upwards towards the head and downwards into the thorax.
(*c*) The contents and boundaries of the thorax to be examined. The several parts of the heart to be observed and the vessels which are connected with it. After removing the contents of the thorax, the sympathetic ganglia and the nerves (splanchnic) leading downwards from them towards the abdomen should be noticed.
(*d*) The abdominal and pelvic viscera to be displayed.
(*e*) The nerves and muscles of one of the hind limbs to be dissected.

Since these dissections form a part of every course of Biology, the student will probably already be familiar with them. If not, they should be made now.

Part 8

BIOLOGY/BOTANY

Editorial Headnote

Emily L. Gregory, 'The Old and New Botany', *Botanical Gazette* (1887), 253–254

Emily Lovira Gregory, 1840–1897, was an American botanist who worked at Bryn Mawr College, the University of Pennsylvania and Barnard College. Trained in Europe, here she writes on the method of teaching botany as a specialism and its relation to the study of biology in general.

33

'THE OLD AND NEW BOTANY'

Emily L. Gregory

Source: *Botanical Gazette* (1887), 253–254

THE OLD AND NEW BOTANY

In the April number of your magazine, the editorial on the methods of teaching botany attracted my attention. While heartily agreeing with the general sentiment expressed, I wish to offer some criticism, or call attention to that aspect of the new method which is considered, in some quarters at least, as the only true biological method.

You close the article with the sentence, "The botanical teaching of the future will consider these, not as two opposing methods, but as complementary, both essential to the rounding out of a botanical course." This implies that at the present time these two methods, the old and the new, may be considered as opposing each other. It seems to me we do not need to wait for the future to teach us that there can be no real opposition between them, for according to their definition in the editorial they refer simply to different departments of the same science. By the old method is meant the teaching of systematic botany, by the new, the teaching of types and the grounding in biological (physiological?) principles. But biological work appears sometimes to be regarded as a science distinct from both botany and zoology, so that one not unfrequently hears of courses being planned in botany, zoology *and* biology, as though botany and zoology failed to recognize plants and animals as living things! Now this is what I do not believe, and it is this monopoly of vital phenomena set up by biologists for biology to which I object.

Now the new method, which is often referred to as the biological one, is supposed to include what is left out in the old, since they complement each other.

While I am unwilling to call it a *method* of teaching, it seems particularly unfortunate to call it the biological method. This word, as I understand it. belongs to the science by virtue of the objects considered being living things, therefore it is illogical to consider any department of botany as entirely unconnected with biology. To explain more fully, classification is based on morphology, or the doctrine of forms, the forms of the organs by which the plant is able to carry on its own existence and to reproduce its kind. It is as idle to consider the vital processes independently of the organs which exhibit them as it would be to ignore the functions of organs by dealing alone with their forms.

DOI: 10.4324/9781003009337-43

The opposition supposed to exist between the old and the new methods is the result of this misuse of the term biological, and it works harm in two ways: First, those wishing to take up the study of botany are deceived by supposing that there is a short, easy, new method, by which they are going to be led straight to the heart of the science without the tedious circumlocution of learning the names of things. It is natural that they should reject and oppose what they consider the old fashioned way. Secondly, this use of the term, biological method, is apt to lead to misunderstanding on the part of students, well educated in other respects, who are not especially interested in the biological sciences. To such the word protoplasm is destined to call up ideas of life-manifestations in which animals and plants either have no part or are inextricably confused.

As long as botanists are willing to suffer the most important part of botanical teaching to be referred to in such vague terms as to effectually disguise its real nature and even mislead educated people into supposing it can only be taught in connection with the science of animal life, so long will they find it difficult to give to botany the rank which it deserves. In France and Germany the study of the vegetable kingdom has been recognized, for some years, as a distinct science, including several departments; the terms, physiological, anatomical, morphological (which includes systematic botany), being used in a similar manner as in connection with zoology. In my judgment, if we were to follow their example and use these or similar terms when speaking of the different departments of botany it would do much toward obviating the two evils referred to. It would certainly help to place the science on a level with other natural sciences in the minds of those not especially interested, and would have some influence, perhaps, in rescuing botany from the general disfavor into which it appears to be falling. When the different parts of botany are not only referred to but taught in the way suggested, so that a general course will include a knowledge of all its departments, and an advanced course, continued study in any one or more of them, there will be no possibility of opposition, except the natural and healthful one of competition between those striving to do their best in their chosen fields.

Part 9

PSYCHOLOGY

Editorial Headnote

Mary Whiton Calkins, *A First Book in Psychology* (New York: MacMillan, 1910), 6–8

Mary Whiton Calkins, 1862–1930, was a pioneer in the development of the psychology laboratory and was the first woman to finish a doctorate in psychology at Harvard, though Harvard did not award the degree because of institutionalised misogyny. Here, in a rudimentary description of methods, she focuses on the place of 'introspection' as a mode of observation unique to psychology.

34

A FIRST BOOK IN PSYCHOLOGY

Mary Whiton Calkins

Source: New York: Macmillan, 1910, 6–8

II. THE METHODS OF PSYCHOLOGY

The methods of psychology are, in general, the two methods of every science: description (that is, analysis and classification) and explanation. But besides these fundamental forms of procedure, every science has certain methods peculiar to itself; and the method which distinguishes psychology is that of introspection. This follows directly from what has been said of the subject-matter of psychology. Its facts are not the common, independent, externalized facts of the physical sciences, but the inner facts, selves, and ideas. To observe the psychic fact one has not, therefore, to sweep the heavens with a telescope, nor to travel about in search of rare geological formations; but one has merely to ask oneself such questions as: "How do I actually feel?" "What do I mean when I say that I perceive, remember, believe?"

The method has obvious advantages. It makes no especial conditions of time and place; it requires no mechanical adjunct; it demands no difficult search for suitable material; at any moment, in all surroundings, with no external outfit, one may study the rich material provided by every imaginable experience. In an extreme sense, all is grist that comes to the psychologist's mill. The apparent facility of introspection is, however, one of its greatest dangers. Nothing seems easier than to render to ourselves a true account of what goes on in our consciousness. We are tempted, therefore, to overlook the need of training in introspection and to minimize its characteristic difficulties. Chief among these is the change which it makes in its own object. To attend to a particular experience actually alters it. If I ask myself in the midst of a hearty laugh, "Just what is this feeling of amusement?" forthwith the feeling has vanished, and a strenuous, serious mood has taken its place. Much the same is true of every form of consciousness. To observe myself perceiving, remembering, or judging is no longer simply to perceive, to remember, and to judge, but to reflect upon perception, memory, and judgment. It is true, therefore, as many psychologists have shown, that introspection is never

DOI: 10.4324/9781003009337-45

of the immediate present, but is rather a case of memory, and subject, therefore, to all the uncertainties of memory.

The verification of our introspection is best secured by an important subsidiary method shared by psychology with many of the physical sciences – the method of experiment. To experiment is to regulate artificially the conditions of phenomena in such wise as to repeat, to isolate, and to vary them at will. In a multitude of ways, therefore, experiment aids scientific observation. Repetition of phenomena insures accuracy of analysis, and makes it possible to verify the results of a single observation; isolation of conditions narrows the object of study, and avoids the distraction of the observer's attention; and, finally, variation of conditions makes it possible to explain a phenomenon exactly, by connecting it with those conditions only which it always accompanies. But because psychic facts differ from physical phenomena in that they can never be repeated or exactly measured, psychological experiment directly concerns itself with the physical stimulation of psychic facts and with the physical reactions to these stimuli. For example, though I cannot measure the vividness of a memory image, I can count the number of repetitions of a series of words which I read aloud to the person on whom I experiment; and I can compare the number of errors he makes in repeating the word-series when he has heard it once only, three times, or five times. In this way I can gain, experimentally, a conclusion about the relation of memory to frequency of experience, and by repeating the experiment many times with the same individual and with others, I may arrive at some trustworthy general conclusion.

INDEX

www.ingramcontent.com/pod-product-compliance
Lightning Source LLC
LaVergne TN
LVHW010630300926
849162LV00001B/106
9780367443740